CLINICAL HEMATOLOGY

MW01070431

MADE RIDICULOUSLY SIMPLE

Stephen Goldberg, M.D.
Professor Emeritus
University of Miami
Miller School of Medicine
Miami, Florida

MEDMASTER

Copyright © 2023 by Medmaster, Inc.

All rights reserved. This book is protected by copyright. No part of it may be reproduced, stored in a retrieval system, or transmitted in any form or by any means (electronic, mechanical, photocopying, recording or otherwise) without written permission from the copyright owner.

ISBN #978-1-935660-47-7

Cover by Matteo Farinella

Made in the United States of America

Published by Medmaster, Inc.
P.O. Box 640028
Miami, FL 33164

Contents

Preface

Hematology encompasses numerous diseases. This book is directed toward the medical, nursing, and PA student as well as the general practitioner, who would like a brief overview of the key and practical clinical aspects of Hematology, with understanding, rather than rote memorization.

I thank Hematologist/Oncologist James Hoffman, MD, Associate Director of the Hematology-Oncology Fellowship Program at the Sylvester Comprehensive Cancer Center, University of Miami Miller School of Medicine, for his insights and contributions, particularly in the chapters on hematologic malignancy. I also thank Phyllis Goldenberg for her excellent proofreading.

Stephen Goldberg, MD

Hematology Made Ridiculously Simple

"For the life of all flesh is its blood."
Leviticus 17:14

1

Introduction

For millennia, through many wars and wounds, humans have come to appreciate the connection of blood to life, but the function of blood remained unknown until relatively recently. It was only in 1628, with the work of William Harvey, that it became known that blood circulates around the human body. We now know that circulation is necessary because a multicellular organism, unlike a single-celled one, requires a circulation to reach the inner body to transport nutrients and remove wastes.

The blood circulation resembles a busy freeway with many drivers going to distant places. The network of arteries, veins, and capillaries stretches an estimated 60,000 miles. The 9-12 pints of blood that the average adult has travels about 12,000 miles/day.

The important passengers in the blood include:

- *Red blood cells* to bring oxygen to the tissues and remove carbon dioxide
- *White blood cells* to fight infection
- *Platelets* and *clotting factors* to plug breaks in the vessels and promote wound healing
- Numerous *plasma** proteins, carbohydrates, lipids, hormones, and electrolytes for the growth and maintenance of the body. Plasma makes up over half of the blood volume (**Figure 1-1**). (**Serum* is plasma minus its clotting factors and is the liquid that remains after blood clots, as in a test tube.)
- *Water* (about 90% of the plasma).
- *Waste products*, e.g. carbon dioxide, urea, and lactic acid
- *Heat*, for regulating the body temperature

Hematology encompasses numerous diseases, and it is easy to get lost in the details of a reference text. This book focuses on seeing the overall clinical picture in a brief, clear manner.

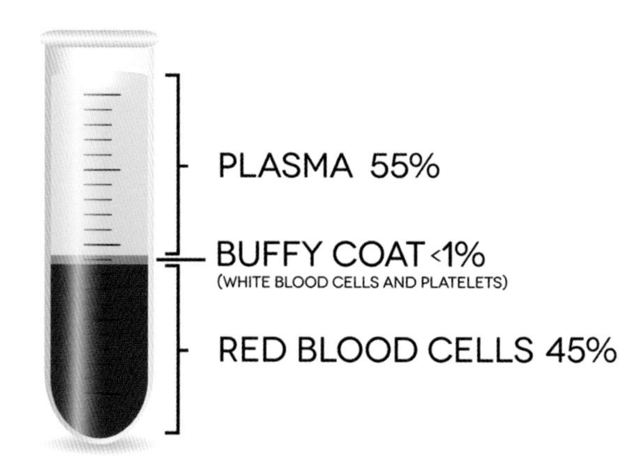

Figure 1-1. *Blood components of unclotted blood. Hematocrit = 45%.*

2

The Cellular Components of the Blood

Figure 2-1 shows the general plan of development of blood cells from a multipotential stem cell in the bone marrow. These cells include:

- *Red blood cells* (RBCs, erythrocytes)
- *White blood cells* (WBCs, leukocytes)
  - *Granulocytes* (neutrophils, basophils, eosinophils, which contain granules)
  - *Monocytes*
  - *Lymphocytes* (T and B lymphocytes, and natural killer cells)
- *Platelets*. Large *megakaryocyte* cells in the marrow break up to form the platelets.

The red cells, white cells, and platelets all originate in the bone marrow, with some exceptions:

- The **liver** is the main site of origin of red cells (**erythrocytes**) in the fetus, and the spleen to a lesser extent. In adults, RBCs can also originate in the liver, spleen and other areas (*extramedullary hematopoiesis*) in certain diseases where the bone marrow does not produce enough red blood cells.
- **T** and **B lymphocytes** are also produced in the **lymph nodes** and **spleen**. T cells (the "T" is for thymus) mature in the **thymus gland**.
- Neutrophils, eosinophils, and basophils are classified as **granulocytes**, since they have visible granules with important functions. Granulocytes only live about 8 hours to a few days in the blood and have to be continually produced. Red cells live longer, about 120 days.

- **Neutrophils** (also called *polymorphonuclear cells, PMN,* because their nuclei normally have 3-5 segments in a variety of shapes (**Figure 2-1**) make up about 40-60% of the WBCs in the blood. They are the first cells to respond when there is tissue damage or infection, particularly bacterial. They migrate to the infection site where they kill bacteria and fungi, and ingest foreign debris. Neutrophil granules release enzymes that kill the invaders or inhibit their growth. In serious infections, the marrow produces numerous neutrophils (**neutrophilia**). Sometimes they are released in an immature state, in which case they have a single curved nucleus, called a **band (Figure 3-19B)**. An increase in bands, termed a *left shift* on the complete blood count (**CBC**), is a sign of infection or inflammation. Having too few neutrophils is termed **neutropenia**.

Generally, but not always, neutrophil number increases in acute processes and in bacterial infections, while lymphocyte number increases in chronic processes and viral infections. Viruses, though, can have elevated lymphocytes in the acute phase, while an elevated neutrophil count can occur in chronic inflammation. Parasites are initially accompanied by a rise in eosinophils. An extremely high WBC count may occur with a malignancy of white blood cells.

- **Eosinophils** release enzymes that kill parasites (e.g. worms) and are involved in allergic responses. They also are active against cancer cells.

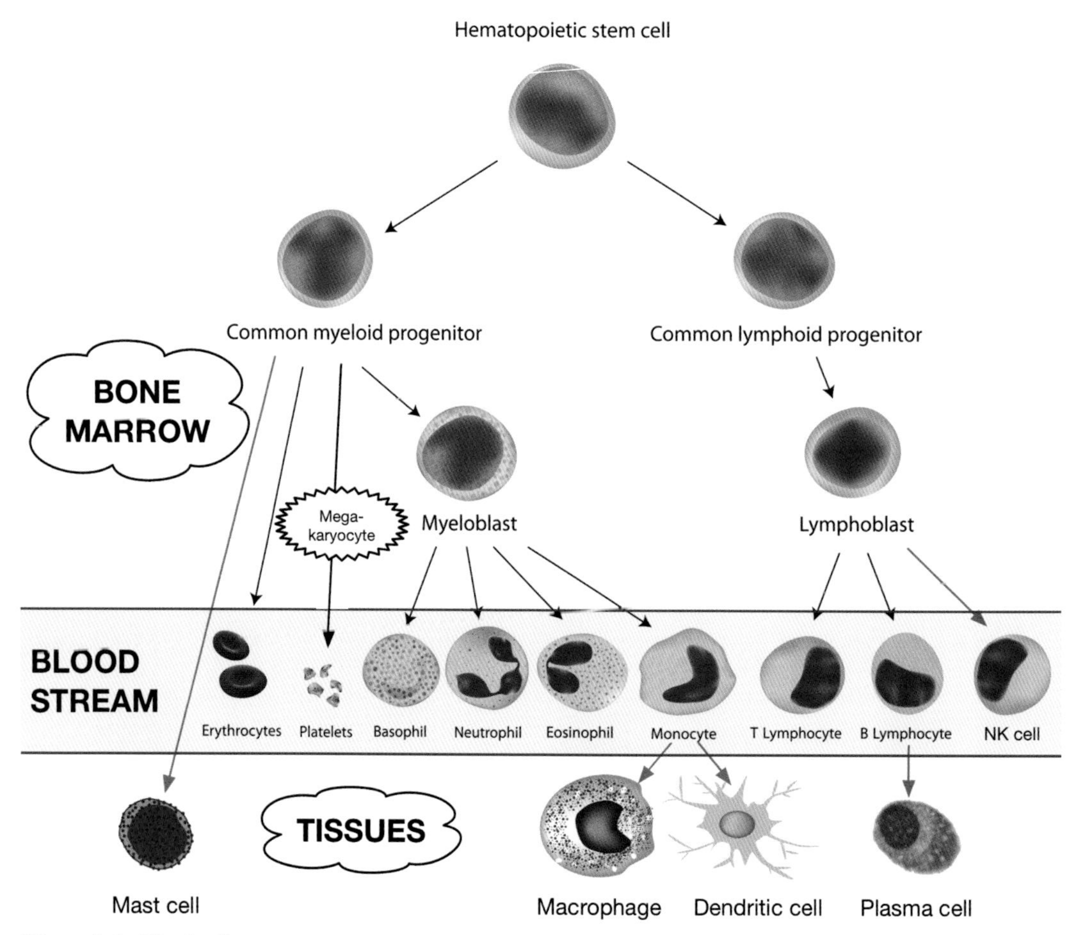

Figure 2-1. Blood cell precursors.

- **Basophils** contribute to the inflammatory response by secreting histamine, which increases blood flow to the tissues by dilating the blood vessels. They also secrete heparin, an anticoagulant that prevents clotting and enables other WBCs to move to the inflammatory site.
- **Monocytes** remain in the blood for only a few hours before migrating into the tissues, where they become **macrophages** and **dendritic cells**, which *phagocytose* (engulf) and digest microbes, dead neutrophils, and other cellular debris. Macrophages and dendritic cells also play more complex roles in the immune system (**Chapter 6**).
- **Lymphocytes** destroy virus-infected cells and foreign invading cells, as well as cancer cells, when lymphocytes recognize those cells as foreign. B lymphocytes develop into **plasma cells**, which reside outside the blood vessels and produce antibodies, which circulate in the blood. The activity of B and T lymphocytes in the immune system is further elaborated on in **Chapter 6**. Lymphocytes may live anywhere from a week to several years. The large range is due to the formation of *memory cells*, which remember past immune reactions and respond quickly on repeat exposure to the foreign substance.
- **Platelets** are fragments in the blood of large **megakaryocyte** cells of the bone marrow. Like red cells, they have no nucleus or mitochondria, but still deliver a powerful punch in starting off blood clotting. They live about 5-9 days.

There are several kinds of related cells that do not circulate in the blood, but inhabit tissues outside the blood vessels:

- **Mast cells**, while originating in the bone marrow, enter the circulation in an immature form and quickly leave the bloodstream to reside in the tissues. They play an important role in the inflammatory response to injury, infection, and allergy.
- **Dendritic cells** and macrophages, as mentioned above, arise from monocytes. They act as phagocytes and also present antigens to T lymphocytes, an important function in the immune system (see **Chapter 6**).
- **Plasma cells** are rarely found in the circulation. They derive from B lymphocytes and reside mostly in lymph nodes and spleen, where they produce antibodies (**Chapter 6**).

3

The Complete Blood Count (CBC)

Sometimes, the first indication of what is wrong when the patient is ill is the routine **Complete Blood Count (CBC)**. (**Figure 3-1**)

Direct visual examination of the blood smear is also useful. It can detect abnormal cell morphologies (**atypical cells**) that are noted more easily by the human eye than by a lab machine.

Red Blood Cell (RBC) Count

In order for body cells to receive enough oxygen, there needs to be enough functional hemoglobin in the blood. **Anemia** is a condition where there is either a decrease in the number of red blood cells or the red cells do not have enough normal-functioning hemoglobin. In **polycythemia**, there are too many red cells.

Figure 3-2 lists clinical conditions that can change the red blood cell (RBC) count.

Hematocrit (Hct) and Hemoglobin (Hgb)

The **hematocrit** (**Hct**) is a percentage ratio of the *volume of red blood cells* compared with the total volume of blood, as measured in an anticoagulated test tube after centrifugation (about 45% in **Figure 1-1**). The **hemoglobin** (**Hgb**) is the *amount of hemoglobin* in the blood as expressed in grams per deciliter (g/dl). In practice, the Hct and Hgb are roughly equivalent in their diagnostic meaning and can be used interchangeably. The conditions that cause an increase or decrease in Hct/Hgb essentially parallel those of increased or decreased RBC count, although it is possible to have an elevated RBC count along with a low Hgb/Hct. For example, in beta-thalassemia trait, the marrow produces a normal or increased number of RBCs. However, the RBCs are small (microcytic), and the amount of hemoglobin in the cells is decreased, manifest as a low Hgb/Hct.

Mnemonic: The approximate rule of 3's. *3 x RBC count = Hgb; 3 x Hgb = Hct*. For example, a normal red cell count of 5,000 corresponds to a Hgb of about 15 and a Hct of about 45. However, this only applies when the RBCs are normal. In thalassemia trait, for instance, the red cell count may be increased, while the Hgb and Hct are decreased.

Mean Corpuscular Volume (MCV)

The average volume of the red blood cell (**MCV**) is a useful criterion in determining the cause of an anemia. **Figure 3-3** lists clinical conditions that can alter the MCV.

Red Cell Distribution Width (RDW)

RDW is a measure of the range of sizes of the red cells, which normally varies between 6-8 micrometers. **Figure 3-4** lists clinical conditions that can cause an altered RDW.

FIGURE 3-1. NORMAL CBC VALUES IN HEMATOLOGY	
RBC	4.7-6.1 M/mcL
WBC	4.8-10.8 K/mcL
Hgb	14.0-18.0 g/dL
Hct	42-52%
MCV	80-100 fL
RDW	11.5-14.5 %
Iron	60-170 mcg/dL
Ferritin	12-300 ng/mL
TIBC	240-450 mcg/dL
Neutrophils	1.8-7.8 K/mcL (40-60%)
Lymphocytes	1.0-4.8 K/mcL (20-40%)
Monocytes	0-0.8 K/mcL (2-8%)
Eosinophils	0-0.45 K/mcL (1-4%)
Basophils	0-0.2 K/mcL (0.5-1%)
Platelets	150-450 K/mcL
MPV	7.4-11.0 fL
PT	11-13.5 seconds
PTT	25-35 seconds
dL = deciliter; **L**= Liter; **mcg** = micrograms; **mcL** = microliter; **ng** = nanograms	
Hct = hematocrit; **Hgb** = hemoglobin; **MCV** = mean corpuscular volume; **MPV** = mean platelet volume; **PT** = prothrombin time; **PTT** = partial thromboplastin time; **RDW** = red cell distribution width	
Note: Normal values vary in different labs and with age. In general, RBC counts tend to be higher in men than in women, while platelet counts tend to be higher in women. People of African ancestry tend to display lower Hgb, WBC, and platelet counts than white persons.	

FIGURE 3-2. RBC COUNT (RBC)

INCREASED RBC COUNT can be due to:
Overproduction of RBCs
- *Hypoxia* (oxygen deficiency, as in moving to a high altitude), cyanotic heart disease, chronic obstructive pulmonary disease, sleep apnea. The stimulus to red cell production is the secretion of *erythropoietin* by the kidney in response to hypoxia.
- *Renal cell carcinoma* that secretes excess erythropoietin.
- *Polycythemia vera*, a cancer of the bone marrow, where the marrow produces too many red blood cells
- Artifact of *dehydration*. The total number of RBCs is normal but appears elevated in a blood sample because of their increased concentration (e.g. in diarrhea, burns).

DECREASED RBC COUNT can be due to:
- ***Decreased production of red blood cells***
  - Poor dietary intake, or *malabsorption*. The necessary ingredients for making RBCs (e.g. iron, folic acid, vitamin B12) are deficient.
  - Bone marrow compromise (e.g. leukemia, aplastic anemia)
  - *Anemia of chronic disease* (iron can't be utilized to create new RBCs although iron stores are normal or high)
- ***Increased loss of red blood cells***
  - *Hemorrhage*
  - *Hemolytic anemia* of numerous causes
  - Burns, infections, frostbite
  - Artifact of *hemodilution* (The total number of red cells may be normal but are diluted with excess body fluid.)

Iron

Iron is an essential ingredient of the heme in hemoglobin (**Figure 4-3**). **Figure 3-5** lists clinical conditions that can change the amount of serum iron.

FIGURE 3-3. MEAN CORPUSCULAR VOLUME (MCV)

INCREASED MCV (increased size of red blood cells) can be due to:
- ***Abnormal red cell production*** in the bone marrow
  - *Megaloblastic anemia* (B12 deficiency, folic acid deficiency).
  - Heavy alcohol use may be accompanied by a decreased intake of B12 and folic acid, or may directly interfere with RBC development.
  - *Hemolysis* (Red cell destruction results in increased numbers of *reticulocytes*, which are incompletely developed red cells released early from the bone marrow in order to deal with the hemolysis and chronic hemorrhage. Reticulocytes are larger than normal RBCs).
  - *Myelodysplasia* (a pre-leukemic bone marrow that releases larger, immature cells).
  - *Drugs*. Drugs that interfere with DNA metabolism and retard normal maturation of red blood cells in the marrow (e.g. methotrexate, azathioprine) may result in a megaloblastic anemia. Certain drugs may also cause hemolysis.
- **Lab artifact**, e.g. cold agglutinins and *rouleaux* formation, where red cells clump and are mistakenly counted as larger cells

DECREASED MCV can be due to:
- *Iron deficiency anemia*, where RBCs are poorly developed, pale (hypochromic) and microcytic
- *Beta-thalassemia trait*, where RBCs are poorly developed, missing part of the hemoglobin protein, and microcytic
- *Lead poisoning* (lead may compete with iron for absorption)
- *Anemia of chronic illness* (iron, while present, is not effectively utilized to form red cells)
- *Congenital sideroblastic anemia* (defective RBC formation through failure of iron to incorporate into heme)

FIGURE 3-4. RED CELL DISTRIBUTION WIDTH (RDW)

INCREASED RDW (anisocytosis = unequal RBC sizes) can be found in:

- *Iron deficiency anemia.* An anemia with a *normal RDW* may suggest thalassemia minor, which can otherwise be confused with iron deficiency anemia.
- *B12 and folate deficiency.* A high RDW can provide a tipoff to a B12 or folate deficiency even before those anemias are detected. High RDW and *high* MCV suggest megaloblastic anemia (e.g. folate/B12 deficiency). High RDW and *low* MCV occur in iron-deficiency anemia.
- *Sickle cell anemia.*
- *Myelofibrosis* (a bone marrow cancer with extensive bone marrow scarring. It disrupts normal RBC production and is associated with a severe anemia.)
- *Hemolytic anemias* (reticulocytes released by the bone marrow are larger than normal RBCs with which they mix).
- *A mixture of a microcytic and macroscopic anemia* in the same patient.
- Conditions where *fragments of blood cells*, *agglutinated blood cells*, or *abnormally shaped blood cells* are counted along with normal cells.
- A *false high RDW* can occur if EDTA (Ethylenediaminetetraacetic acid) anticoagulated blood is used instead of citrated blood (may be due to clumping).

Not every anemia has an increased RDW.

NORMAL RDW can be found in:

- *Anemia of chronic disease*
- *Hereditary spherocytosis*
- *Acute blood loss* (no time for reticulocytosis)
- *Aplastic anemia* (bone marrow failure – including failure to produce reticulocytes)
- Some cases of *thalassemia minor*

FIGURE 3-5. SERUM IRON

INCREASED SERUM IRON can be due to:

- ***Increased iron intake***
  - Dietary
  - Iron poisoning
  - Hemochromatosis, a disease of excess iron absorption
- ***Excess release of iron from cells***
  - Hemolytic anemia
  - Blood transfusions
  - Lead toxicity (lead displaces stores of iron)
  - Liver disease (e.g. hepatitis, which releases iron stores into the blood. Iron builds up because the normal body does not have a good way to excrete excess iron.)

DECREASED SERUM IRON can be due to:

- ***Poor dietary intake of iron***
- ***Malabsorption of iron***
  - Prolonged achlorhydria (acid is required to release iron from food)
  - Sprue/celiac disease
  - Excess starch/clay eating (may inhibit iron absorption)
  - Short bowel syndrome (Iron is normally absorbed in the small intestinal duodenum and upper jejunum.)
- ***Excess iron loss***
  - Hemorrhage (GI bleeding, uterine bleeding, hematuria)
  - Late pregnancy (where the fetus depletes the mother's iron supply)

Ferritin

Iron is mostly stored in **ferritin**, a protein complex mainly stored in the liver (in *hepatocytes*) but also in macrophages in the bone marrow, spleen, duodenum, skeletal muscle and other areas. There is very little ferritin in the blood, but, importantly, enough to be tested. **Figure 3-6** lists clinical conditions that can change ferritin levels.

FIGURE 3-6. SERUM FERRITIN
INCREASED SERUM FERRITIN can be due to: • ***Increased iron storage need***, which requires more ferritin to store the excess iron in an iron overload. - *Hemochromatosis (hemosiderosis)*. Storage is needed for the excess absorbed iron. - *Hemolysis* (red cell destruction) releases more iron to store. - *Megaloblastic anemia*. Iron is not being incorporated into blood cells and needs to be stored. - *Iron poisoning.* - *Hyperthyroidism* (may increase ferritin synthesis). • ***Inflammation***. Inflammation generates an immune response that prevents iron from being utilized, but is stored instead, so more ferritin is needed to store the available iron. - Chronic infection - Certain cancers - Autoimmune diseases - Kidney and liver disease **DECREASED SERUM FERRITIN** can be due to: • ***Decreased serum iron.*** Since there is less iron, there is less need to produce ferritin to store it. - Iron deficiency anemia (less iron requires less ferritin to store it) - Dialysis (can reduce iron during the process) • ***Decreased serum protein***. Ferritin, a protein, decreases with protein deficiency.

A low serum ferritin level is usually diagnostic of iron deficiency anemia (IDA). Ferritin is a storage form of iron and it is the *storage of iron that depletes first* before the anemia is manifest. Ferritin decline is an earlier sign of iron deficiency than other parameters of the CBC, such as red cell morphology, Hct, Hgb or serum iron (**Figure 3-7**).

Total Iron Binding Capacity (TIBC)

*Trans*ferrin is the main protein that *trans*ports iron from one place to another in the blood, for use or storage. **Total Iron Binding Capacity (TIBC)**, sometimes called **transferrin iron-binding capacity**, measures the maximum amount of iron that can be carried by the transferrin in a blood sample. The less transferrin there is, the less iron that it can bind. TIBC is an indirect measure of the amount of blood transferrin and is less expensive to test than directly measuring transferrin.

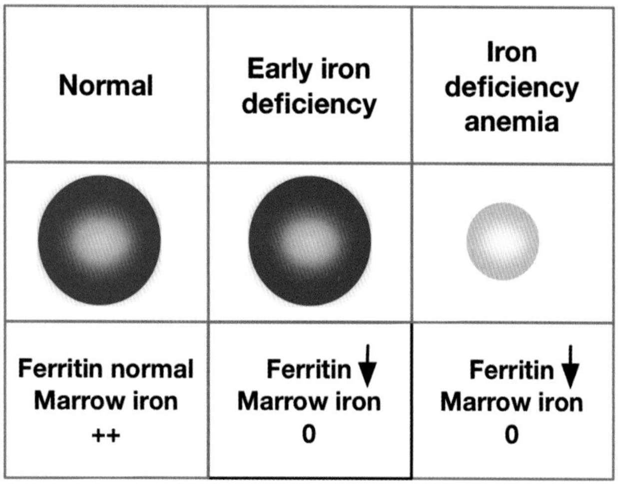

Normal	Early iron deficiency	Iron deficiency anemia
Ferritin normal Marrow iron ++	Ferritin ↓ Marrow iron 0	Ferritin ↓ Marrow iron 0

Figure 3-7. Stages of iron deficiency. Storage forms of iron are the first to decline.

- *Iron deficiency* increases transferrin synthesis (and TIBC) in the body's effort to scrounge around for more iron in the blood for delivery to the tissues. As the amount of transferrin increases in iron deficiency anemia, the percentage of *iron saturation* in transferrin decreases.
- *Increased serum iron* (iron overload, as in hemochromatosis, where the body absorbs too much iron) reduces transferrin synthesis (and TIBC), since the body tries to decrease the overload of iron in the blood. The body tries to keep a balance in the amount of iron that is available. The less transferrin, the less iron it can bind. So TIBC decreases with iron overload, while transferrin saturation increases.
- *Anemia of chronic disease* (ACD, as in chronic infection, autoimmune diseases, cancer, kidney disease) reduces transferrin synthesis. This is a preventative way the body can reduce the availability of iron as a nutrient for bacteria and tumor cells. In ACD, iron is kept in storage and not utilized. So the ferritin increases to deal with the excess iron storage, while the transferrin (TIBC) decreases.

In general, regardless of the disease, whenever the ferritin level is up, transferrin (TIBC) level is down, and vice-versa (**Figures 3-8** and **3-9A**). If you already know the ferritin is up or down, you can predict whether the TIBC level is down or up respectively.

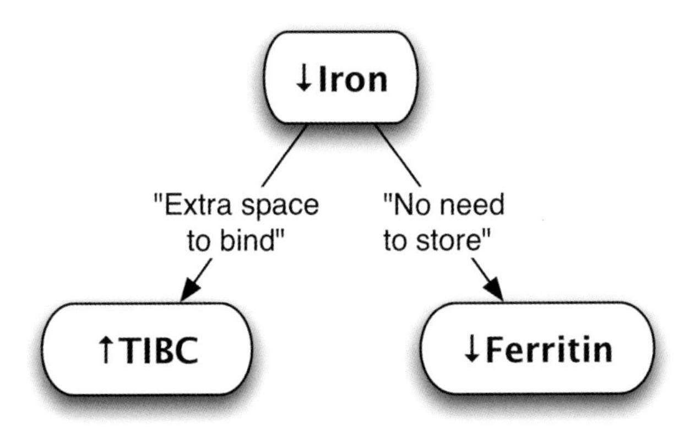

FERRITIN AND TRANSFERRIN (TIBC) IN IRON DEFICIENCY ANEMIA

Figure 3-8. From Berkowitz, A. Clinical Pathophysiology Made Ridiculously Simple, Medmaster.

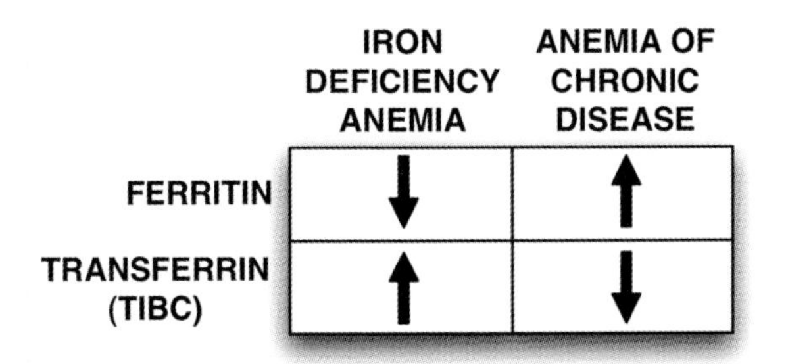

FERRITIN AND TRANSFERRIN (TIBC) IN ANEMIA OF CHRONIC DISEASE

Figure 3-9A. From Berkowitz, A. Clinical Pathophysiology Made Ridiculously Simple, Medmaster.

The diagnosis of the type of anemia is not always clear when there is a coinciding mixture of iron deficiency (e.g. from bleeding) with an anemia of chronic disease (e.g. cancer). The former lowers the ferritin, while the latter raises it.

A bone marrow biopsy will reveal absent iron staining in iron deficiency anemia, but not in anemia of chronic disease. A bone marrow biopsy, though, is invasive, and it may be best, if possible, to rely on other aspects of the history, physical, and lab to determine the diagnosis.

Figure 3-9B lists clinical conditions that can cause a change in TIBC.

Reticulocyte Count

Increased numbers of **reticulocytes** appear in the blood when the bone marrow rapidly produces many new red cells due to erythropoietin action and releases immature RBC reticulocytes into the bloodstream.

Figure 3-10 lists conditions that can change a reticulocyte count.

Figure 3-11 summarizes CBC lab values in anemia.

White Blood Cell (WBC) Count

An elevated white blood cell count is common with infections, inflammation, and leukemia. In leukemia, the count can be extremely high (100,000-400,000 per microliter, in contrast with the normal range of about 4,000-11,000).

The type of white cell that increases depends largely on the kind of infection and the kind of leukemia. The white cell count may actually show a decline in overwhelming

FIGURE 3-9B. TOTAL IRON BINDING CAPACITY (TIBC)

INCREASED TIBC (transferrin) can be due to:
- ***Iron deficiency***. There is more transferrin, so its capacity to bind iron is high in iron deficiency anemia.
- ***Other***
  - Estrogen excess (pregnancy, birth control pills) leads to increased transferrin synthesis independent of body iron level.

DECREASED TIBC can be due to:
- ***Increased serum iron***. There is less need to produce more transferrin.
  - Hemolytic anemia (iron overload)
  - Increased dietary iron intake
  - Hemochromatosis, a disease of excess iron absorption
  - Pernicious anemia (Iron is not incorporated into red cells; instead it is stored in ferritin.)

- ***Decreased transferrin synthesis*** can be due to:
  - Cirrhosis (Transferrin is normally synthesized in the liver, which is damaged.)
  - Malnutrition/hypoproteinemia (Transferrin is a protein that may be decreased in malnutrition.)
  - *Atransferrinemia*, a rare hereditary disease where there is an absence of transferrin

- ***Chronic inflammation***
  - Anemia of chronic disease (chronic infection, autoimmune diseases, cancer, kidney disease). There is poor utilization of available iron, which remains stored, coupled with decreased production of transferrin.

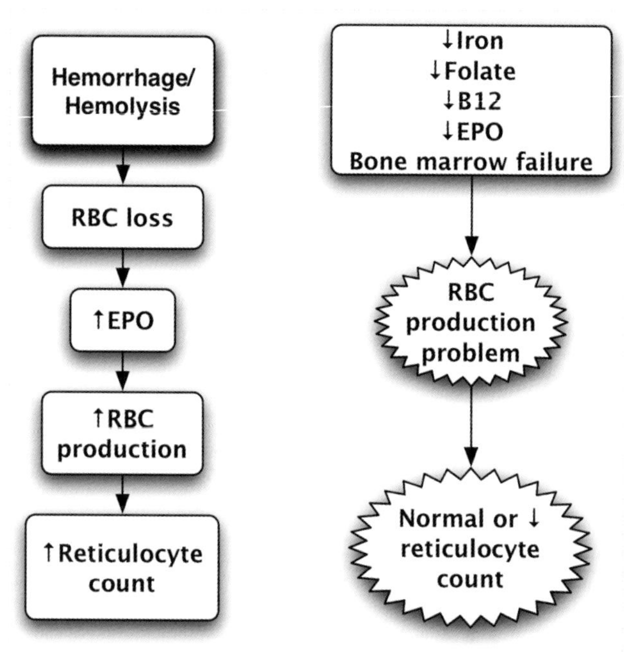

Figure 3-10. *From Berkowitz, A. Clinical Pathophysiology Made Ridiculously Simple, Medmaster. EPO= erythropoetin.*

FIGURE 3-10. RETICULOCYTE COUNT
INCREASED RETICULOCYTE COUNT can be found in: • ***Hemorrhage*** • ***Response to hemolysis*** • ***Move to high altitude (anoxia)*** due to increased erythropoietin production • ***Response to treatment of anemias*** (corrected with iron, B12, or folate) • ***Erythropoietin-secreting tumors*** (e.g. adrenal cell carcinoma, hepatocellular carcinoma, hemangioblastoma)
DECREASED RETICULOCYTE COUNT may indicate: • ***Deficiency of B12, folic acid, or iron.*** Reticulocyte count is low because there aren't enough ingredients to form new red cells, including reticulocytes. • ***Anemia of chronic disease*** (decreased production of RBCs and reticulocytes) • ***Bone marrow failure*** (aplastic anemia, tumor, certain drugs, radiation, infection)

bacterial infections that use up neutrophils faster than they can be replenished, particularly in the elderly. WBC numbers may all decline in bone marrow failure, as a side effect of certain drugs (e.g. in chemotherapy), radiation, and as a response to hypersplenism (the spleen, among other things, removes white cells), cachexia, and as a congenital hereditary condition. All the white blood cells may show an artifactual rise in the setting of dehydration.

Neutrophil Count

Neutrophils characteristically predominate as a normal response to bacterial infections, although they may also predominate in certain viral and other infections (**Figure 3-12**).

An increased neutrophil count may occur in response to corticosteroids because:

1. Steroids promote **demarginalization** of neutrophils, namely the tendency of neutrophils to detach from the blood vessel endothelium and mix with the other neutrophils floating in the blood. *Epinephrine* also promotes demarginalization. There is no actual increase in the neutrophils, but more are counted.
2. The neutrophils transmigrate less through the vessel wall and remain in the bloodstream.
3. Glucocorticoids and epinephrine induce the release of non-segmented bands of neutrophils from the blood marrow. Exercise also promotes leukocytosis, perhaps as a reaction to epinephrine release.

Lymphocyte Count

Lymphocytes tend to predominate in viral infections, but there are certain bacterial infections where lymphocytes are prominent. A rise in lymphocytes tends to appear later than neutrophils in inflammation. **Figure 3-13** lists factors that can cause a change in lymphocyte count.

Lymphopenia occurs when the body doesn't make enough lymphocytes, or the lymphocytes the body makes are destroyed or sequestered in the spleen or lymph nodes.

Eosinophil Count

Eosinophils increase in allergies, in certain autoimmune disease, and in response to parasitic worm infections. They are prominent in a variety of tumors. **Figure 3-14** lists factors that can cause a change in eosinophil count.

Decreased eosinophil count does not usually indicate a problem unless all the other white cell counts are low, too.

Monocyte Count

Monocytes are prominent in certain infections as well as certain autoimmune diseases, cancers, and other conditions.

Figure 3-15 lists conditions that can change the monocyte count.

FIGURE 3-11. LAB VALUES IN ANEMIA								
	RBC	Hgb/Hct	MCV	RDW	Iron	Ferritin	TIBC	Retics
Iron deficiency	↓	↓	↓	↑	↓	↓	↑	↓
B12/folate deficiency	↓	↓	↑	↑	↓	↑	↓	↓
Anemia of chronic disease	↓	↓	↓	nl	↓	↑	↓	↓
Hemolytic anemia	↓	↓	↑	↑	↑	↑	↓	↑
Sickle cell	↓	↓	↑/↓	↑	↑	↑	↓	↑
Thalassemia	nl/↑	↓	↓	↑	↑	↑/nl	↓	↑/nl
RBC enzyme deficiency	↓	↓	↑	↑	↑	↑	↓	↑
Porphyria	↓	↓	nl	↑	↑	↑	↓	↑
Sideroblastic anemia	↓	↓	↓	↑	↑/nl	↑	↓/nl	↓
Aplastic anemia	↓	↓	↑	nl	↑	↑	↓	↓
Hemochromatosis	nl	nl/↑	nl	(nl)	↑	↑	↓	↑
Iron overload					↑	↑	↓	

FIGURE 3-12. NEUTROPHIL COUNT

INCREASED NEUTROPHIL COUNT (neutrophilia) may occur in:

- ***Acute infections***
  - Bacterial
    - Cocci (staphylococcus, pneumococcus, streptococcus, gonococcus, meningococcus)
    - Bacilli (E. coli, pseudomonas, actinomyces)
  - Certain fungi (Coccidioides immitis)
  - Spirochetes
  - Viruses (chicken pox, herpes zoster, rabies, polio, smallpox)
  - Rickettsia
  - Parasites (liver fluke)
- ***Inflammation/necrosis/trauma***
  - Myocardial infarction
  - Gout
  - Thyroiditis
  - Burns
  - Postoperative period
  - Acute glomerulonephritis
  - Rheumatic fever
  - Hypersensitivity reactions
  - Collagen vascular diseases
- ***Myeloproliferative disorders***
  - Acute myeloid leukemia
  - Chronic myeloid leukemia
  - Myelofibrosis (neutrophil count may also be normal or decreased in some)
  - Myeloid metaplasia
- ***Metabolic disorders***
  - Diabetic ketoacidosis
  - Uremia
  - Preeclampsia
  - Gout

Toxic chemicals and drugs

  - Lead, mercury, digitalis, phenacetin, quinidine, turpentine, insect venom

- ***Acute hemorrhage***: The neutrophilia may result from inflammation and/or release of corticosteroids and epinephrine, which raise the WBC count.
- ***Rapidly growing tumors***: Neutrophilia may be due to tumor necrosis factor-alpha, which induces inflammation.

(Continued)

Figure 3-12 (Continued)

FIGURE 3-12. NEUTROPHIL COUNT
• ***Exercise***: may relate to epinephrine release, which raises WBC count • ***Corticosteroids*** (e.g. Cushing syndrome): In infection, though, corticosteroids may also prevent the expected rise in in WBC count. • ***Hereditary*** (***leukocyte adhesion deficiency*** – leukocytes can't leave blood vessel to migrate to the site of infection to form pus; multiple infections) • ***Idiopathic***
DECREASED NEUTROPHIL COUNT (**neutropenia**) may result from: • ***Certain infections*** (including overwhelming bacterial infections, particularly in the elderly) - Typhoid - Brucellosis - Measles - Influenza - Epstein-Barr virus - Cytomegalovirus - Viral hepatitis - HIV-1 - Toxoplasmosis - Brucellosis - Tuberculosis - Malaria - Dengue fever - Babesiosis - Rickettsial infection • ***Bone marrow failure*** - Aplastic anemia - Hematologic malignancy (leukemia, lymphoma, myeloma, myelodysplasia) - Radiation treatment - Metastatic tumor to bone marrow - Myelofibrosis - Granulomatous infection • ***Immune-mediated neutropenia*** - Drug origin - Autoimmune origin - Rheumatoid arthritis - Lupus - Sjogren syndrome - Hodgkin lymphoma - Autoimmune hepatitis - Thymoma - Goodpasture disease (anti-neutrophil antibodies) - Wegener granulomatosis (anti-neutrophil antibodies) • ***Hypersplenism*** (the spleen destroys white cells at an abnormal rate) • ***Cachexia*** • ***Hereditary*** - *Chediak-Higashi syndrome* (immune deficiency with recurrent infections; reduced pigment in skin and eyes) - *Cyclic neutropenia* (recurrent episodes of reduced neutrophils in blood) - *Cartilage-hair hypoplasia syndrome* (immune deficiency with recurrent infections; skeletal abnormalities and sparse hair) - *Dyskeratosis congenita* (bone marrow failure; rashes; white patches in mouth, misshapen fingernails) - *Infantile genetic agranulocytosis* (*Kostmann syndrome* – lack of neutrophils; susceptibility to infections) - *Lazy leukocyte syndrome* (neutropenia with abnormal neutrophil motility) - *Myelokathexis* (neutrophils fail to enter the bloodstream) - *Schwachman-Diamond syndrome* (bone marrow failure; poor growth) - *Reticular dysgenesis* (immunodeficiency; poor WBC development; deafness)

FIGURE 3-13. LYMPHOCYTE COUNT

INCREASED LYMPHOCYTE count (lymphocytosis) may occur in:

- ***Infections***
  - Viral infections
    - Mononucleosis
    - Primary infection with HIV
    - Viral hepatitis
    - Mumps
    - Chicken pox
    - Influenza
    - Rubella
    - Measles
    - Infectious lymphocytosis
    - Coxsackie virus B2
    - Poliovirus
    - Enterovirus
  - Bacterial infections
    - Chronic bacterial infections
    - Pertussis (Bordetella pertussis)
    - Cat scratch fever (Bartonella henselae)
    - Brucellosis
    - Tuberculosis
  - Other infections
    - Syphilis
    - Malaria
- ***Hypersensitivity reactions***
- ***Chronic inflammation***
- ***Lymphocytic leukemia***
  - Chronic lymphocytic leukemia
  - Acute lymphocytic leukemia
- **Stress** (lymphocytosis may be due to epinephrine)
  - Trauma
  - Cardiac emergencies
  - Status epilepticus
- **Other**
  - Smoking
  - Post-splenectomy (less destruction of lymphocytes)
  - Radiation

DECREASED LYMPHOCYTE COUNT (lymphopenia) may occur with:

- ***Viral Infections***
  - HIV
  - Ebola
  - Epstein-Barr virus
  - Advanced viral hepatitis
  - Influenza
- ***Bacterial infections***
  - Tuberculosis
  - Typhoid fever
  - Sepsis
- ***Fungi/Parasites***
  - Histoplasmosis
  - Malaria
- ***Drugs***
  - Chemotherapy/radiotherapy
  - Immunosuppressants (steroids)
- ***Autoimmune disorders***
  - Rheumatoid arthritis
  - Lupus
  - Aplastic anemia
  - Myasthenia gravis
  - Chronic kidney disease
  - Sarcoidosis
  - Multiple sclerosis
- ***Hematologic cancer and other blood diseases***
  - Leukemias and metastatic tumors of the bone marrow
  - Aplastic anemia
- ***Hereditary***
  - Hematopoietic stem cell aplasia
  - Ataxia-telangiectasia (immunodeficiency; decreased coordination; spider veins)
  - Cartilage-hair hypoplasia syndrome (immune deficiency with recurrent infections; skeletal abnormalities and sparse hair)
  - Idiopathic CD4+ lymphocytopenia
  - Thymoma immunodeficiency
  - Severe combined immunodeficiency syndrome
  - Wiskott-Aldrich syndrome (immune deficiency; eczema; low platelet count)
  - Adenosine deaminase deficiency (immune deficiency; infections)
  - Purine nucleoside phosphorylase deficiency (immune deficiency; infections; neurologic symptoms; autoimmune disease)
  - DiGeorge anomaly (immunodeficiency; congenital anomalies)
- ***Malnutrition***
- ***Stress***

Basophil Count

Basophils increase in certain infections and inflammations, leukemias, hemolytic anemias, following splenectomy, and certain endocrine disorders.

Figure 3-16 lists factors that can cause a change in basophil count.

Platelet Count

Platelets are critical elements in the initiation of blood clotting.

Figure 3-17 lists factors that can cause a change in platelet count.

FIGURE 3-14. EOSINOPHIL COUNT

INCREASED EOSINOPHIL COUNT may occur in:

- ***Allergy***
  - Asthma
  - Allergies
  - Autoimmune disorders (atopic dermatitis, Crohn disease, ulcerative colitis, and other autoimmune diseases)
- ***Parasitic worm infections***
  - Ascariasis
  - Schistosomiasis
  - Trichinosis
  - Visceral larva migrans
  - Strongyloidiasis (roundworm)
  - Gnathostomiasis (roundworm)
  - Fascioliaisis (liver fluke)
  - Paragonimiasis (lung fluke)
- ***Infections***
  - Scarlet fever (group A strep)
  - Epstein-Barr virus
  - Aspergillus
  - HIV
  - Tuberculosis
- ***Tumors***
  - Gastric carcinoma
  - Lung carcinoma
  - Lymphoma
    - Hodgkin lymphoma
    - Non-Hodgkin lymphoma
    - Human T-cell lymphotropic virus (HTLV-1)
    - Adult T-cell leukemia/lymphoma (ATLL)
    - Eosinophilic leukemia
- ***Idiopathic***

DECREASED EOSINOPHIL COUNT may occur in:

- Alcohol intoxication
- Excess cortisol production (e.g. Cushing syndrome) or Rx with corticosteroids
- Morning as opposed to evening

FIGURE 3-15. MONOCYTE COUNT

INCREASED BLOOD MONOCYTES can be found in:

- ***Infections***
  - Bacterial
    - Tuberculosis
    - Brucellosis
    - Subacute bacterial endocarditis
  - Spirochetes
    - Syphilis
  - Viruses
    - Infectious mononucleosis
  - Protozoa and rickettsia
    - Malaria
    - Rocky Mountain spotted fever
    - Kala-azar
- ***Neoplasia***
  - Hodgkin disease
  - Monocytic leukemia
  - Myleoproliferative diseases
- ***Autoimmune disease***
  - Lupus
  - Rheumatoid arthritis
  - Ulcerative colitis and other inflammatory bowel diseases
- ***Other***
  - Sarcoid
  - Lipid storage diseases
  - Dehydration (artifactually elevates all leukocytes due to hemoconcentration)

DECREASED BLOOD MONOCYTES

A low monocyte count can be due to anything that decreases the overall white cell count, with accompanying neutropenia and lymphopenia.

Atypical Cells

Apart from being able to view differences in cell size and number, the blood smear can also assess atypical-looking cells that provide clues to the diagnosis (**Figure 3-18** – red cell morphologies; **Figure 3-19** – white cell morphologies).

Atypical Red Blood Cells (Figure 3-18)

Normal red blood cell (Figure 3-18A). Non-nucleated, central pale area without inclusions.

Microcytic hypochromic red blood cell (Figure 3-18B). Small, pale red blood cell most commonly found in iron deficiency

Target cells (Figure 3-18C) (increased cell surface/Hgb ratio due to either excess cell surface area or decreased hemoglobin content). The RBCs do not carry oxygen efficiently. Seen in thalassemia, iron deficiency anemia, sickle cell disease, hemoglobin C and D diseases, and after splenectomy (altered RBCs are no longer removed by the spleen). Liver disease may alter the lipid synthesis needed for the cell membrane structure.

Nucleated RBCs (Figure 3-18D) in severe hemolysis with premature release of the altered form. Normally, 1% or less of RBCs should be nucleated.

Basophilic stippling (Figure 3-18E). Numerous basophilic granules of aggregated RNA, ribosomes, and mitochondria; indicates disturbed erythropoiesis (e.g. thalassemia, hemolytic anemia, sickle cell, myelodysplastic syndrome, leukemia, or heavy metal poisoning, such as lead).

Reticulocytes (Figure 3-18F). Immature non-nucleated red cells released early from the bone marrow to deal with hemolysis and chronic hemorrhage. Reticulocytes are larger than normal RBCs. Staining shows residual RNA and mitochondria.

Spur cells (acanthocytes) (Figure 3-18G). RBCs with thorny projections in cirrhosis, pancreatitis and other conditions. Cells have poor deformability.

FIGURE 3-16. BASOPHIL COUNT

INCREASED BASOPHIL COUNT may be found in:
- ***Myeloproliferative disorders***
  - Chronic myeloid leukemia
  - Polycythemia vera
  - Primary myelofibrosis
  - Essential thrombocythemia
  - Myelodysplastic syndrome
  - Systemic mastocytosis
  - Hypereosinophilic syndrome
- ***Inflammation***
  - Inflammatory bowel disease
  - Rheumatoid arthritis
  - Asthma
  - Chronic sinusitis
  - Chronic dermatitis/psoriasis
  - Hashimoto thyroiditis
- ***Allergies***
  - Food allergies
  - Drug allergies
  - Hay fever
  - Allergic rhinitis
- ***Infections***
  - Chickenpox
  - Tuberculosis
  - Parasitic infections
- ***Endocrine***
  - Hypothyroidism
  - Increased estrogen

DECREASED BASOPHILS may be found with:
- Hypersensitivity reactions (e.g. allergic urticaria)
- Hyperthyroidism
- Corticosteroids
- Lupus
- Pregnancy
- Hereditary absence of basophils

FIGURE 3-17. PLATELET COUNT

INCREASED PLATELET COUNT may be found in:
- ***Excess production of platelets***
  - Essential thrombocytosis
    - Acute hemorrhage
    - Acute infection
    - Arthritis and other chronic inflammations
    - Chronic myelogenous leukemia
    - Lymphoma
    - Certain tumors of the colon
    - Polycythemia vera
    - Myeloid metaplasia
    - Iron-deficiency anemia (Iron deficiency induces a general attempt by the marrow to produce cells. Platelet production occurs, since it is not dependent on the presence of iron.)
    - Osteoporosis
    - Stress
    - Surgery
- ***Decreased elimination of platelets***
  - Post-splenectomy (The spleen normally helps remove platelets)
- ***Normal physiological difference***

DECREASED PLATELET COUNT may occur with:
- ***Defective production of platelets***
  - Leukemia and other myelofibrosis diseases (The marrow is replaced, including the capacity to produce platelets.)
  - Cancer chemotherapy (destroys bone marrow)
  - Pernicious anemia (B12 is needed for platelet production.)
  - Wiskott-Aldrich syndrome (immune deficiency; bleeding; eczema)
  - Bernard-Soulier disease (platelet defect; bleeding)
- ***Increased removal of platelets***
  - Hypersplenism
  - Hemorrhage (Platelets are lost during bleeding.)
  - Immune thrombocytopenic purpura (ITP)
  - Cirrhosis with hypersplenism (The spleen removes platelets.)
  - Posttransfusion purpura
  - Neonatal alloimmune thrombocytopenia (materal antibodies against fetal platelet antigens). Healthy women may have an inconsequential mild thrombocytopenia during pregnancy.
  - Drug-induced thrombocytopenia (antibiotics, cardiac drugs, anti-rheumatics, heparin)
  - Disseminated intravascular coagulation (DIC) (platelets are used up)
  - Thrombotic thrombocytopenic purpura (There is a large amount of abnormal von Willebrand protein, which is involved in platelet adhesion.)
  - Hemolytic uremic syndrome (kidney failure; low red cells and platelets)
  - Infection (can sometimes result in thrombocytopenia)
  - Artifact of platelet clumping in EDTA (as opposed to sodium citrate) test tube anticoagulation

They arise from membrane defects in protein or lipids (e.g. apolipoprotein A-II deficiency). Seen in *abetalipoproteinemia*. Not like Burr cells, where spikes are shorter and at regular intervals on the cell membrane.

Burr cells (Figure 3-18H) (echinocytes) in kidney failure (uremia) and in hemolytic anemia – short, evenly spaced spikes like sea urchins. May also be an artifact of anticoagulation. Spikes are shorter and more numerous than in spur cells. Mechanism: outer lipid bilayer expands outward relative to the inner lipid bilayer.

Howell-Jolly bodies (Figure 3-18I). Little fragments of RBC nucleus that the spleen normally removes. May be seen in any condition where there is loss of splenic function.

Sickle cells (Figure 3-18J) in sickle cell anemia

Elliptocytes (Figure 3-18K) in hereditary elliptocytosis (problem with membrane protein), iron

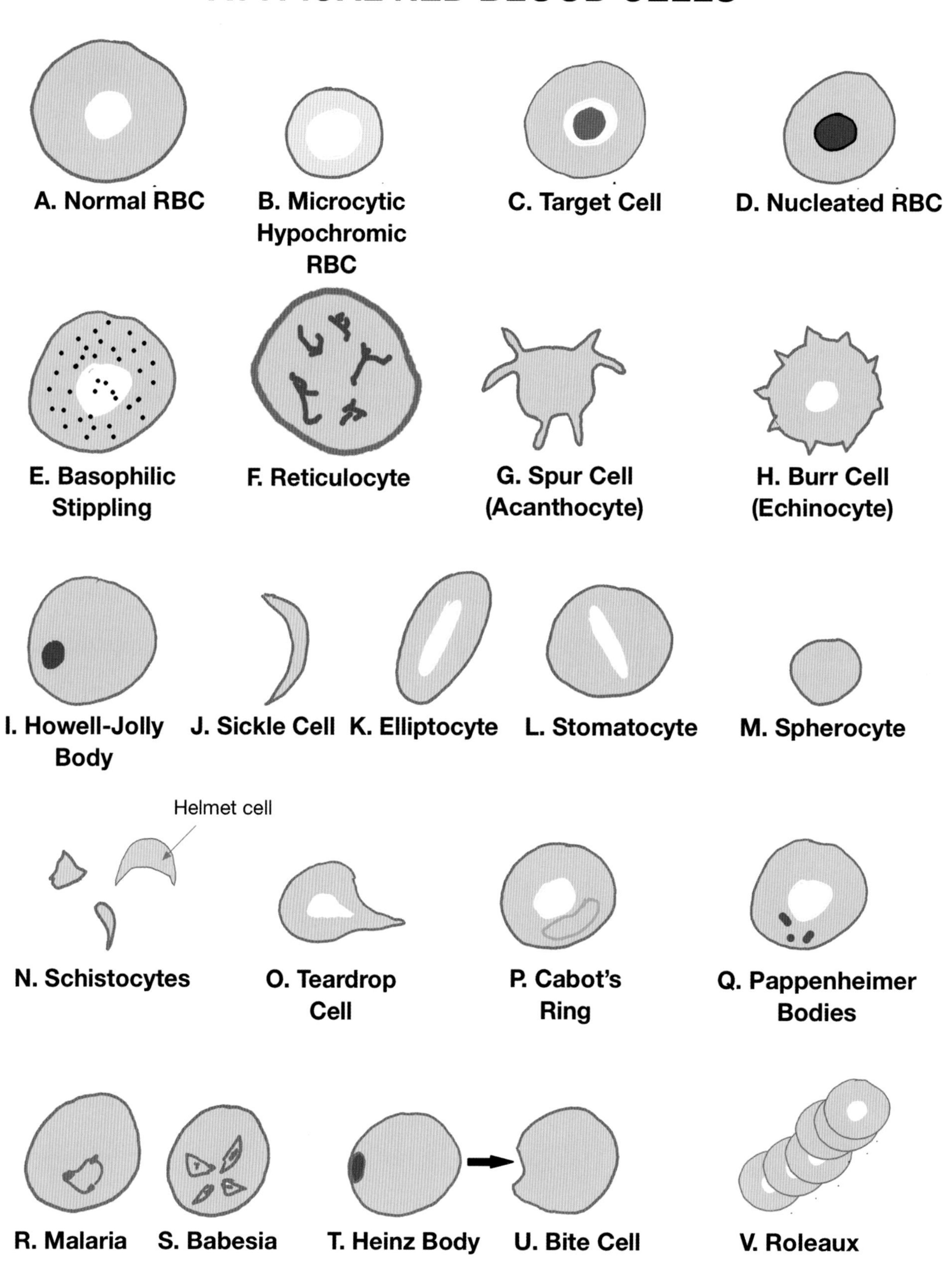
ATYPICAL RED BLOOD CELLS
A. Normal RBC
B. Microcytic Hypochromic RBC
C. Target Cell
D. Nucleated RBC
E. Basophilic Stippling
F. Reticulocyte
G. Spur Cell (Acanthocyte)
H. Burr Cell (Echinocyte)
I. Howell-Jolly Body
J. Sickle Cell
K. Elliptocyte
L. Stomatocyte
M. Spherocyte
Helmet cell
N. Schistocytes
O. Teardrop Cell
P. Cabot's Ring
Q. Pappenheimer Bodies
R. Malaria
S. Babesia
T. Heinz Body
U. Bite Cell
V. Roleaux

Figure 3-18.

deficiency anemia, thalassemia, megaloblastic anemia, myelodysplastic syndrome.

Stomatocytes (Figure 3-18L). *Stomatocytosis* is a rare condition where the red cells have a slit-like central zone instead of the usual round central pallor. The cell is fragile and subject to hemolysis. The condition may be hereditary or the result of excess alcohol ingestion, in which case it may disappear after about 2 weeks of alcohol withdrawal.

Spherocytes (Figure 3-18M). The cell is sphere-shaped rather than its usual flat, round doughnut shape. Found in hereditary spherocytosis and immunologically caused *extravascular hemolysis*. The cell is fragile due to a defect in the RBC membrane with increased permeability to sodium and water, which causes the cell to swell and burst.

Schistocytes (**Figure 3-18N**). Jagged fragments (triangular, **helmet** or comma-shaped) of RBCs seen in *intravascular hemolysis*. Those cells with a few very large protuberances are called **horn cells**.

Teardrop cells (**dacryocytes**) **(Figure 3-18O)** are found in bone marrow fibrosis and myelproliferative diseases. May be caused by mechanical squeezing out from the bone marrow as a result of bone marrow infiltration.

Cabot's rings (**Figure 3-18P**) are rare threadlike, looping strands (possibly of microtubules) found in cases of abnormal RBC production in B12 deficiency and lead poisoning.

Pappenheimer bodies (Figure 3-18Q) are cytoplasmic iron granules found in sideroblastic anemia.

Parasites (Figure 3-18R,S). Malaria (**Figure 3-18R**), a single-celled parasitic organism, presents as irregular inclusions on Giemsa stain, often ring-shaped. **Babesia** (**Figure 3-18S**) is a parasite spread by a tick bite. Symptoms may resemble those in malaria. Its configuration in some of the red cells resembles a Maltese cross.

Heinz bodies (Figure 3-18T) are denatured, clumped hemoglobin (seen in G6PD deficiency and other RBC enzyme defects). When spleen macrophages bite them out, they are then called **bite cells (Figure 3-18U)**.

Rouleaux formation (Figure 3-18V) (like a stack of coins) in connective tissue diseases, diabetes, multiple myeloma (immunoglobulin-coated RBCs cling together), antibiotic allergy, and diabetic retinopathy. Occurs where plasma protein (particularly fibrinogen and globulins) is high.

Atypical White Blood Cells (Figure 3-19)

Normal neutrophil (Figure 3-19A). The nucleus contains 3-5 segments.

Bands (Figure 3-19B). Immature neutrophils in serious infections (>15% bands = "left shift")

Hypersegmented neutrophils (Figure 3-19C) (6 or more lobes or > 3% of neutrophils with at least 5 nuclear lobes) in B12 and folate deficiency, and myeloproliferative disorders. Reason unclear. **Bilobed neutrophils** are found in *Pelger-Huet anomaly,* a rare, inherited, benign condition.

Toxic granulation in neutrophils (Figure 3-19D). Large granules in the cytoplasm of segmented and band neutrophils in blood are found in severe infections. Indicates phagocytosis and lysosomal activity.

Bright green inclusions (Figure 3-19E) in liver failure (lipid inclusions) (poor prognosis)

Atypical (reactive) lymphocytes (Figure 3-19F) are large, distorted lymphocytes with large irregularly shaped nuclei. Found in certain infections, particularly viral (e.g. mononucleosis, cytomegalovirus, hepatitis B), immune reactions to transplantation and immunization, autoimmune diseases, cancer, and drug reactions.

Blasts (Figure 3-19G). Immature WBCs (**myeloblasts, metamyelocytes, promyelocytes, myelocytes**, or in the case of the lymphatic system, **lymphoblasts** or **prolymphocytes**). Suspicious for leukemia.

Smudge cells (**basket cells**) **(Figure 3-19H)** are characteristic of chronic lymphocytic leukemia. They are the remnants of abnormally fragile, ruptured lymphocytes and are without nuclear or cytoplasmic structure.

Atypical Platelets

Platelets too large (increased mean platelet volume, MPV) are seen in *immune thrombocytopenic purpura*, *Bernard-Soulier disease*, and *white platelet syndrome*, an autosomal dominant disorder with thrombocytopenia and prolonged bleeding.

Platelets too small (MPV) are found in thrombocytopenia arising from an aplastic anemia, as well as in *Wiskott-Aldrich syndrome* (an immune deficiency syndrome, with susceptibility to infections and bleeding).

Hypogranular platelets are found in *white platelet syndrome* (thrombocytopenia; increased bleeding tendency; enlarged platelet size; abnormal internal platelet structure; decreased responsiveness to stimuli to platelet aggregation).

Nonspecific Tests

The **erythrocyte sedimentation rate (ESR)** measures how fast red cells settle to the bottom in an anticoagulated tube. Normal ESR is about 1-20 mm/hr, males in the lower range. ESR may be elevated in anemia, infection, pregnancy, aging, inflammation, autoimmune diseases, some kidney diseases and cancers. (High fibrinogen and other blood proteins

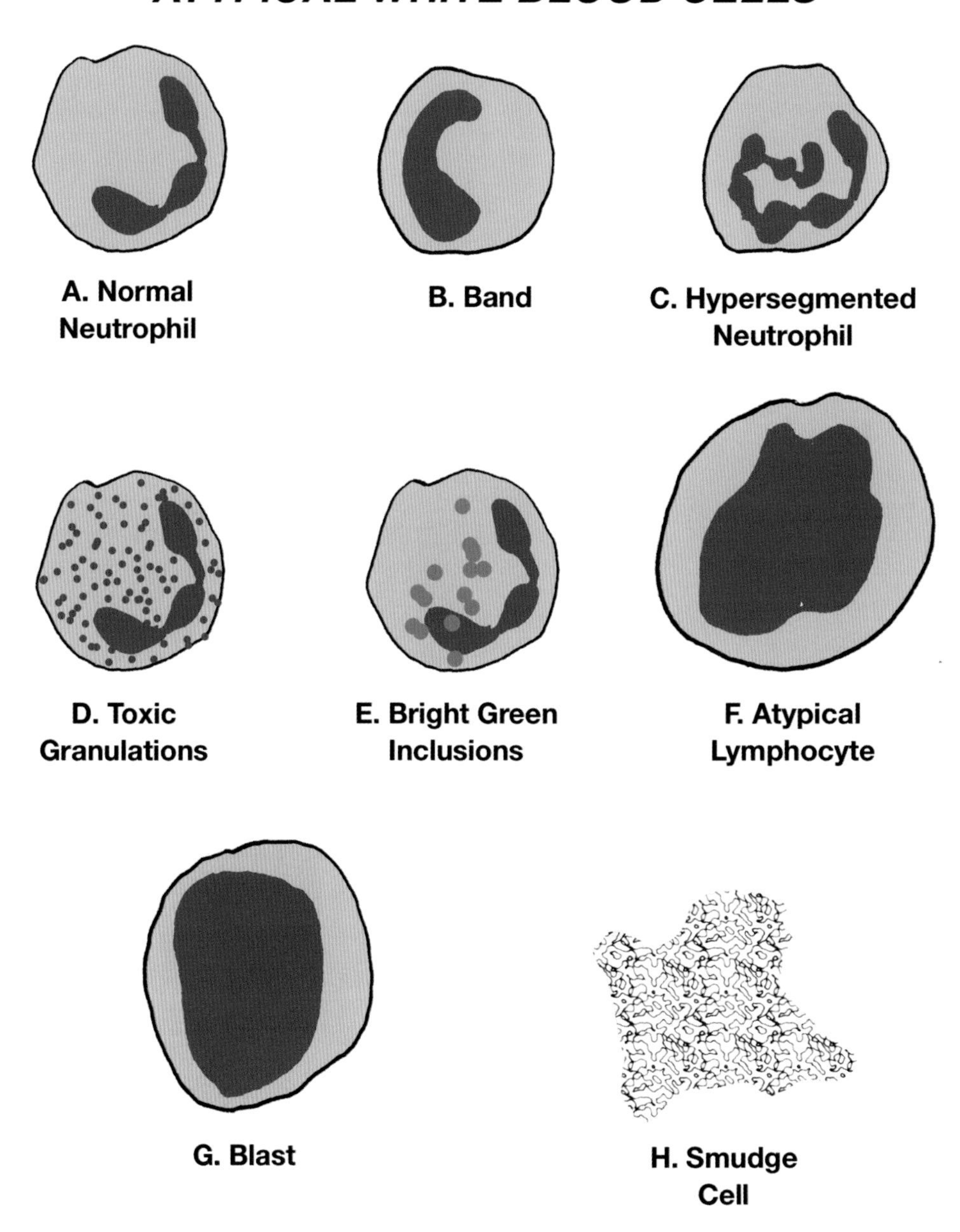

Figure 3-19.

cause red cells to stick together and settle faster.) The ESR is a rather nonspecific test that generally just tells you that "something is wrong" and may merit further investigation, particularly if it is more than 100 mm/hr. The ESR can be used to monitor the progress of the diagnosed disease.

C-reactive protein is a plasma protein that rises (>3mg/L) in response to inflammation. While not pointing to what or where the inflammation is, its level can be used to monitor the response to medication or the progression of the disease in question.

4

Red Blood Cells

Anemia

Anemia is a condition where the blood does not carry enough oxygen, either because of decreased production of normal red blood cells, bleeding, or hemolysis.

General Symptoms and Signs of Anemia

General symptoms of anemia may include:

- fatigue
- weakness
- skin pallor
- lightheadedness
- shortness of breath
- irregular heartbeats
- chest pain
- headaches
- cold hands and feet

Since these are general findings in all anemias, the text, to avoid repetition, will simply refer to them as *general anemic symptoms* when describing the various anemias. Individual anemias may have other symptoms, depending on the condition.

Causes of Anemia

Figures 4-1A and 4-1B summarize the main causes of anemia.

Hemoglobin is constructed of globin *protein* and *iron-containing heme* (**Figure 4-3**). Defects in construction of any part of hemoglobin can cause anemia:

- In **sickle cell anemia** and **thalassemia**, it is the **protein** (globin) portion of hemoglobin that is defective. There are 4 globin portions to hemoglobin: 2 alpha and 2 beta globin chains. Defects may involve one or more of them.
- In **porphyria**, it is the **heme** molecule in the hemoglobin that is improperly constructed.
- In **sideroblastic anemia,** there is a defect in the incorporation of **iron** into heme during red cell production.

History in Anemia

As in many areas of medicine, elements of the history often provides key clues:

- A history of previous blood studies (including those of previous rejections for blood donation) helps estimate how long the anemia has been present
- The family history (including jaundice, gallstones, splenectomy, bleeding disorders, abnormal hemoglobin)
- Medications and potentially toxic agents
- History of blood loss, including an increase in menstrual loss (number of tampons/pads used/mo), black stools, GI complaints; change in bowel habits

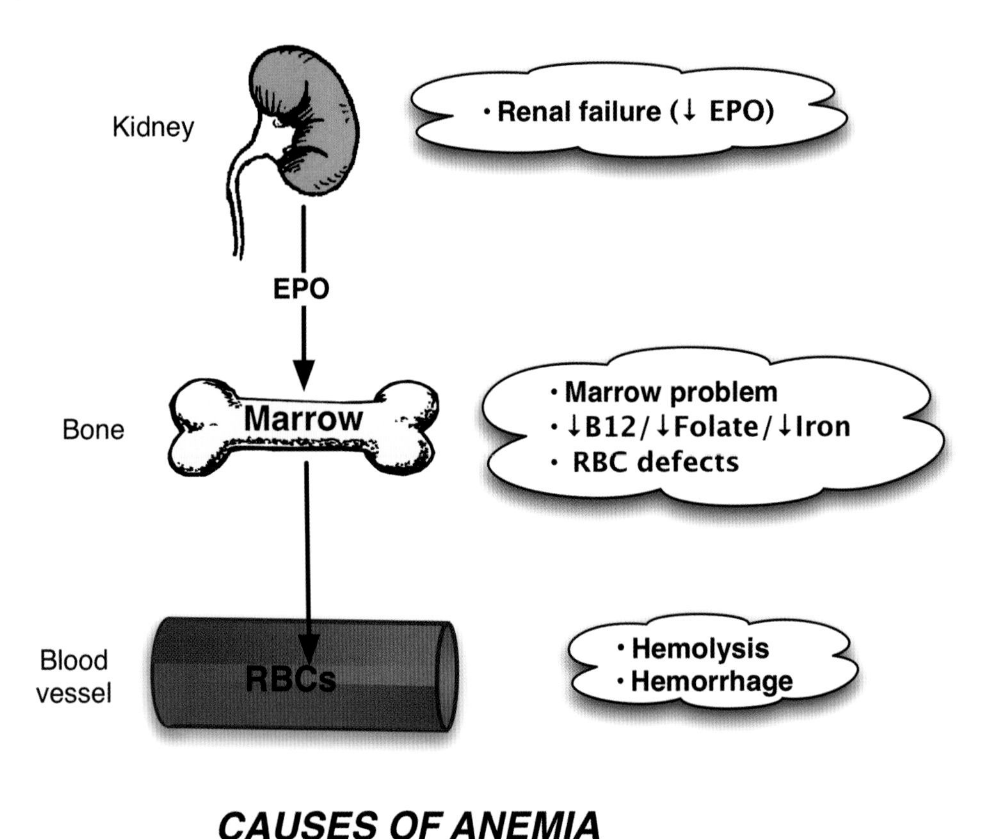

Figure 4-1A. From Berkowitz, A. Clinical Pathophysiology Made Ridiculously Simple, Medmaster.

FIGURE 4-1B. CAUSES OF ANEMIA
DECREASED PRODUCTION OF RED BLOOD CELLS • ***Renal failure***. The kidney does not produce erythropoietin, which is necessary to stimulate red blood cell production in the bone marrow. • ***Bone marrow compromise*** by cancer or other diseases - In *aplastic anemia*, the marrow as a whole is suppressed, commonly by radiation or drugs, but occasionally aplastic anemia is hereditary. In aplastic anemia, red cells as well as other blood cells do not develop in the bone marrow (*pancytopenia*). - *Dietary insufficiency* of ingredients necessary to form red blood cells (e.g. iron, vitamin B12, folic acid)
INCREASED LOSS OF RED BLOOD CELLS - *Hemorrhage* - *Hemolytic anemia* -- *Immune diseases* that destroy RBCs (e.g. autoimmune hemolytic anemia) -- *Defective red cells* (**Figure 4-2**) that render the cell susceptible to hemolysis --- Defective RBC *metabolic processes* for energy production (e.g. G6PD deficiency, pyruvate kinase deficiency) --- Defective red cell *membrane* (e.g. spherocytosis, elliptocytosis, paroxysmal nocturnal hemoglobinuria) --- Defective red cell hemoglobin (e.g. sickle cell, thalassemia, porphyria) -- Mechanical trauma of red cells passing through artificial heart valves or through microthrombi

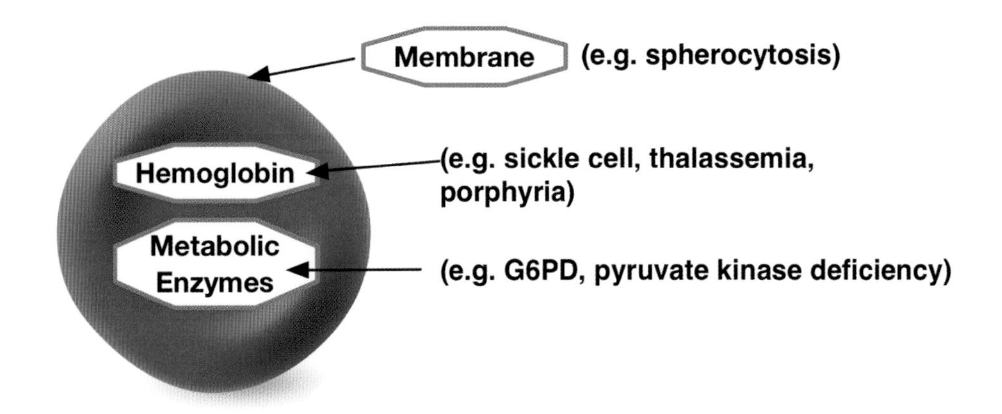

Figure 4-2.

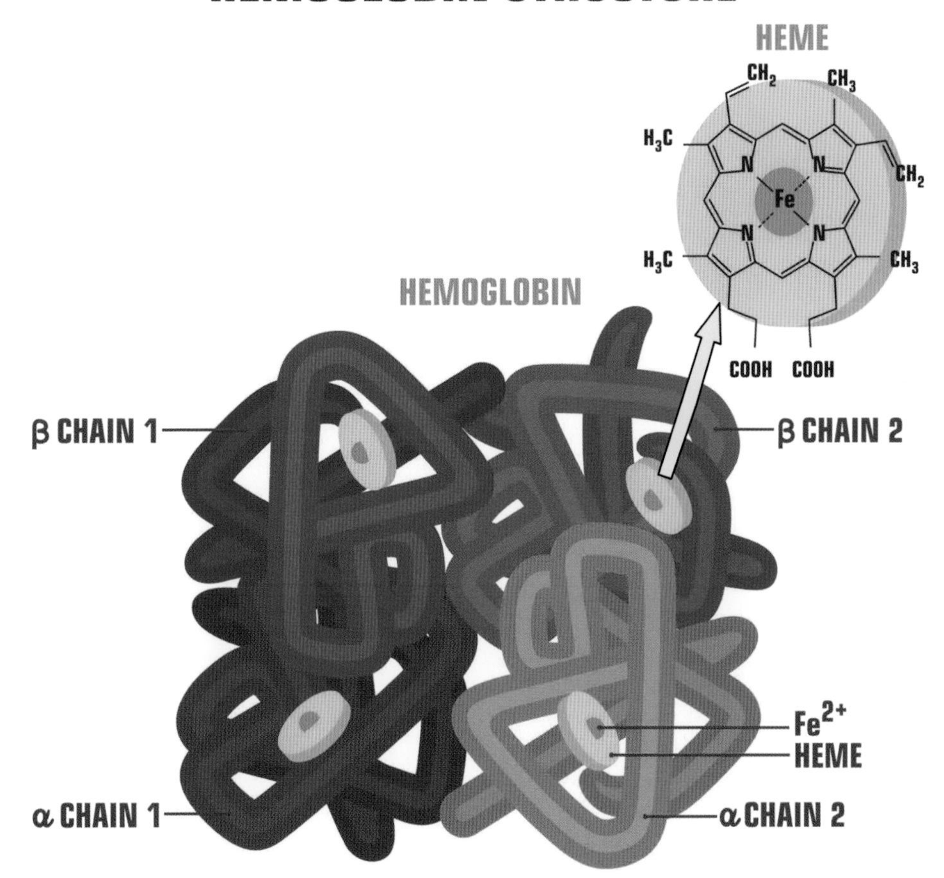

Figure 4-3.

- Change in urine color—e.g. pink/red/brown in *urinary tract bleeding* (hematuria) and *hemolytic anemia* (with jaundice); dark urine in *porphyria*; dark in morning in *paroxysmal nocturnal hemoglobinuria*. In *march hemoglobinuria*, there is mechanical damage to RBCs appearing after marching or running.
- Diet. In iron deficiency anemia, some people eat unusual things of non-nutritive value (**pica**), such as clay, chalk, cigarette butts, soil, paper, or ice.
- *Weight gain* if there is an underactive thyroid. Iron deficiency anemia impairs thyroid metabolism. *Weight loss* despite good nutrition if there is malabsorption.
- *B12 deficiency*: Loss of balance (proprioception defect), burning tongue, early graying of the hair, paresthesias (burning, tingling of skin)
- *Glossitis* (beefy-red sore tongue) and *cheilosis* (cheilitis; inflamed cracks at mouth corners) in iron, folate, or vitamin B12 deficiency
- *Floating stools* – Fatty stools (steatorrhea) from malabsorption, may coincide with B12, intrinsic factor deficiency.
- *Vitamin C deficiency* results in reduced iron absorption and also increases the chance of bleeding.
- *Cold intolerance* – may indicate hypothyroidism, lupus, paroxysmal cold hemoglobinuria, certain macroglobulinemias.

Physical Exam in Anemia

- Is there an appearance of malnutrition?
- Pallor of the conjunctiva, nailbed, and palmar crease
- Iron deficiency anemia: brittle fingernails, *koilonychia* (spoon nails), impotence, leg cramps, *dysphagia* (difficulty swallowing from esophageal webs in *Plummer-Vinson syndrome*, which occurs in long-term iron deficiency)
- Skin: pallor; palmar erythema; dry, brittle hair; puffy face; thinning of the lateral aspects of the eyebrows; dry, itchy skin; jaundice from hemolysis
- Bleeding under the skin (purpura, petechiae in thrombocytopenia)—may mean that other cells besides red cells are defective.
- Glossitis (inflammation of tongue) and angular cheilitis (inflammation of the corners of the mouth) can occur with iron, folate, or B12 deficiency.
- Conjunctiva/sclera: pallor, jaundice, petechiae
- Lymph node enlargement – lymphoma may cause anemia.
- Bilateral edema may indicate coinciding cardiac/renal/hepatic diseases.
- Unilateral edema may indicate lymphatic obstruction from malignancy.
- Hepatomegaly/splenomegaly may occur in a number of anemias.
- Rectal/pelvic: Is there blood in the stool? A tumor may be the cause of the anemia.
- Neurologic exam: position/vibration affected in pernicious anemia.

Iron Deficiency Anemia

Causes of Iron Deficiency Anemia

Iron deficiency anemia may be caused by:

- ***Poor dietary intake of iron***
- ***Malabsorption of iron***
  - Prolonged achlorhydria (acid is required to release iron from food)
  - Sprue/celiac disease
  - Excess starch/clay eating (may inhibit iron absorption)
  - Short bowel syndrome (Iron is absorbed in the small intestinal duodenum and upper jejunum.)
- ***Excess iron loss***
  - Hemorrhage (GI bleeding, uterine bleeding, hematuria)
  - Late pregnancy (where the fetus depletes the mother's iron supply)

Diagnosis of Iron Deficiency Anemia

Apart from the general symptoms of anemia, there may be other clues in the history and physical exam that are more specific to iron deficiency:

- Brittle fingernails, koilonychia (spoon nails)
- Eating unusual things of non-nutritive value (pica), such as clay, chalk, cigarette butts, soil, paper, ice
- Excessive bleeding
- Glossitis (inflammation of the tongue); angular cheilitis; pallor of the conjunctiva, skin, nailbed and palmar crease.

The *laboratory findings* further narrow the differential diagnosis:

- A low **serum iron** may be found in iron deficiency anemia, but in the early stages, serum iron may be normal since there is still enough iron in the blood even though the bone marrow iron stores have been depleted. An acute bleed may not initially show anemia, despite the loss of iron (**Figure 3-7**).
- Low blood **ferritin**. Blood ferritin is a more sensitive indicator of iron deficiency anemia than is serum iron. Iron is mostly stored until it is needed in ferritin, a protein complex mainly stored in the liver but also in the bone marrow, spleen, duodenum, skeletal muscle and other areas. There is very little ferritin in the blood, but enough to be

tested. With iron depletion there is less need for ferritin, and blood ferritin levels *decline* early in the disease, even before serum iron decreases.

- **Total Iron Binding Capacity (TIBC)** (transferrin) increases. Transferrin increase is a reaction of the body to low iron, in an attempt to increase iron transport as much as possible. The more transferrin, the more iron binding capacity (TIBC) the blood has (and the less saturation with iron that it has). In general, a decrease in ferritin is always accompanied by an increase in transferrin (TIBC) and vice versa in *any* of the anemias (**Figure 3-11**).
- The **red blood cell count** (**RBC**) is low in advanced iron deficiency anemia since there is not enough iron to form RBCs. Similarly, the **hematocrit** and **hemoglobin** are reduced.
- The **mean corpuscular volume** (**MCV**) is small (**microcytic**), and the red cells are pale (**hypochromic**) in iron deficiency anemia, as cells do not form properly.
- **Red cell distribution width** (**RDW**) is increased, meaning there is a significantly greater *range* of RBC sizes (**anisocytosis**).
- If iron deficiency is suspected, stool exam for blood, and GI imaging/endoscopy may be indicated.
- If hemolysis is suspected, examine the blood for increased reticulocyte count, indirect bilirubinemia, increased LDH (from RBC destruction), and low haptoglobin (which binds free hemoglobin from hemolysis).

Treatment of Iron Deficiency Anemia

- If the anemia is due to bleeding, the source of the bleeding should be located and corrected, surgically if necessary.
- Iron supplements (commonly ferrous sulphate) and/or dietary change. Supplements may need to be given for 6 months or so to replenish the iron stores and correct the anemia. If the response is inadequate to oral therapy, it is important to be sure the diagnosis is correct and there are no additional causes for the anemia, e.g. B12/folate deficiency, malabsorption, malignancy, or patient noncompliance.
- Treatment may also be given intravenously when there are high iron requirements, e.g. in continual GI bleeding.

B12/Folate Deficiency (Megaloblastic) Anemia

Both **vitamin B12** and **folate** (folic acid) are important in the production of red blood cells. Both are absorbed in the GI tract. **Intrinsic factor**, a glycoprotein produced by the parietal cells of the stomach, is necessary to facilitate the absorption of vitamin B12 in the ileum of the small intestine. A deficiency in B12, folate, or intrinsic factor may cause anemia.

Causes of B12/Folate Deficiency Anemia

Folate and/or B12 deficiency can be caused by:

1. **Poor diet**. Although B12 is not produced by the human body, it is found in animal products like meat, eggs, and shellfish. B12 stores last for years, so dietary deficiency is uncommon (but can occur in a vegan diet). Folate storage is poor, though, lasting only about 4 months, and is more commonly associated with nutritional deficiency, even though folate is common in many foods (leafy green vegetables, citrus fruits, beans, bread, cereals, rice).
2. **Malabsorption**, e.g. tropical sprue; celiac disease; small intestinal resection; Crohn disease; tapeworm.
3. **Excess loss of folate** in:
   - Pregnancy (excess folate utilization in DNA synthesis and breast feeding)
   - Malignant disease
   - Inflammatory diseases
   - Liver disease. There may be resulting dietary insufficiency, poor folate storage, loss of folate in the urine, bleeding, or hemolysis.
   - Drugs that lower folate levels (e.g. anticonvulsants, oral contraceptives, cholesterol-lowering bile acid sequestrants, e.g. cholestyramine, methotrexate)
   - Alcoholism (through inadequate diet, malabsorption, and increased folate excretion)

Intrinsic factor deficiency (and hence B12 deficiency) can be caused by:

- Antibodies to intrinsic factor or parietal cells (pernicious anemia)
- Gene mutation resulting in absent or dysfunctional intrinsic factor
- Gastrectomy

Diagnosis of B12/Folate Deficiency Anemia

Apart from the usual anemic symptoms, patients with B12 or folate deficiency may experience:

- Glossitis (sore, red tongue)
- Numbness/tingling in the hands or feet
- For B12 deficiency (but not folate deficiency), the patient may have balance/gait difficulty and decreased vibratory sense in the lower extremities due to spinal cord posterior column degeneration.
- Mental confusion

- Psychiatric symptoms (e.g. depression, panic attacks, insomnia)
- Jaundice (may be due to breakdown of poorly formed, fragile red cells)

Since many of the symptoms in B12 and folate deficiency are similar, folate and B12 blood levels are commonly measured in the same workup.

Laboratory findings in B12/folate deficiency anemia may include:

- Enlarged (megaloblastic) red blood cells
- Hypersegmented nuclei of neutrophils (**Figure 3-19C**)
- Low serum B12 and folate levels in B12 and folate deficiency respectively, but there is variability in the specificity and sensitivity of these tests.
- Elevated *homocysteine* blood levels in both B12 and folate deficiency, because B12 and folic acid are both necessary to convert homocysteine into methionine. Homocysteine therefore backs up in the blood with either deficiency.
- Elevated *methylmalonic acid* in serum or urine in B12 deficiency, not in folate deficiency. The reason is that B12 (not folate) is necessary to convert methylmalonic acid to succinyl-CoA, so methylmalonic acid backs up and increases with B-12 deficiency alone.
- Antibodies to intrinsic factor
- Antiparietal cell antibodies. Stomach parietal cells produce both hydrochloric acid and intrinsic factor. When parietal cells are damaged by the antibodies, this results in both low stomach acid (*achlorhydria*), low intrinsic factor, and pernicious anemia.
- Serum gastrin levels. *Gastrin* is a hormone that stimulates the secretion of gastric acid (HCl) by the parietal cells of the stomach. Gastrin rises as a feedback response to the achlorhydria of pernicious anemia and is a useful marker of B12 deficiency.
- Assessing gastric pH is also useful in confirming pernicious anemia.
- GI consult to rule out disease of the stomach or intestines
- The neurologic exam is particularly important, since B12 deficiency is associated with significant problems with balance and vibratory sense due to damage to the posterior columns of the spinal cord. These symptoms are generally not present in folate deficiency and are not improved with folate therapy alone. The patient with B12 deficiency may have these early neurologic symptoms without an anemia. B12 or folate deficiency should be considered in the differential diagnosis of memory loss, depression, and other behavioral changes, which can occur in either condition and are potentially treatable.
- The response of the patient to treatment with either B12 or folate replacement therapy is useful in making the diagnosis. Treatment of B12 deficiency may be oral, but if intestinal absorption is at the root of the deficiency, the B12 may need to be given by intramuscular injection or under the skin.

Classically, a test known as the **Schilling test** was used to diagnose pernicious anemia; however, this test is rarely performed now. It involves having the patient orally ingest radioactive B12. If the patient excretes the radioactive B12 in the urine, it indicates normal B12 absorption. If not excreted, it indicates a problem with B12 absorption. If the malabsorption is corrected with the addition of intrinsic factor (IF) to the radioactive B12, it suggests that a lack of IF was the cause of the malabsorption and B12 deficiency, rather than primary malabsorption.

The problem with the Schilling test is a lack of available radioactive B12 and intrinsic factor, and the need to collect a 24-hour urine, which, if not accurate, may lead to misleading results. Also, false positives can arise with decreased renal function. And there are other diagnostic tests.

Treatment of B12/Folate Deficiency Anemia

- B12 deficiency is treated with B12, either orally, or when malabsorption is at the root of the problem, by intramuscular injections of B12. Overdosing with B12 does not occur because B12 is water-soluble and any excess is excreted in the urine. This contrasts with excess iron, which the body cannot remove and can result in iron overload.
- Folic acid deficiency is treated with folic acid supplements and/or dietary improvement. Folic acid will not improve the neurologic problems of B12 deficiency (balance and vibration deficits), which may continue to worsen if not treated with B12, so B12 deficiency must be ruled out.
- Vitamin B12 may be given as prophylactic therapy about every 3 months to patients who are strict vegetarians or have undergone ileal resection. Folic acid is given routinely in pregnancy because of the increased need for folate by the fetus, who otherwise may have increased susceptibility to developing neural tube defects.

Anemia of Chronic Disease (ACD)

In **anemia of chronic disease** (e.g. conditions where there is chronic infection, autoimmune disease, cancer, or kidney disease; also called anemia of inflammation) there is poor utilization of available iron.

ACD may have different mechanisms:

- A dysregulation of iron metabolism is thought to be induced by excess **hepcidin**, a protein that inhibits the release of iron from macrophages and also inhibits the absorption of iron from the GI tract. Iron remains trapped in macrophages and is poorly absorbed from the gastrointestinal tract and therefore cannot be used for hemoglobin synthesis.
- Impaired and reduced production of and/or response to **erythropoietin (EPO)**, particularly in kidney disease (the kidney normally produces erythropoietin).
- Decrease in the life span of red cells. This may be an autoimmune phenomenon; the result of infection; renal disease; or cancer that releases substances that damage immature red blood cells.

Diagnosis of Anemia of Chronic Disease

The patient's chief complaint may be the usual symptoms of anemia. The history and physical exam are particularly important. Are there signs of chronic inflammation or a history of autoimmune disease, cancer, or kidney disease?

Laboratory findings provide further clues:

In ACD, iron is stored excessively and not utilized in making red cells.

- The **ferritin** the iron is stored in *increases* to store all the iron rather than leaving it free to circulate. This a way in which the body can protect against supplying iron nutrient to enemy bacteria or cancer cells. **Transferrin** is reduced as a further way of keeping iron out of the blood. There is also decreased iron absorption in the GI tract.
- The Total Iron Binding Capacity (**TIBC**) is decreased in ACD because there is less **transferrin.**
- **Serum iron** is low in both iron deficiency anemia (IDE) and anemia of chronic disease. In IDE it's because the body as a whole doesn't have enough iron. In ACD, it's because the iron is stored rather than used or circulated.
- Bone marrow aspiration staining for iron can distinguish IDE from ACD. There is no iron staining in IDE, but staining is present in ACD.
- Serum creatinine may be elevated in ICD due to renal insufficiency.

In all kinds of anemia, the total iron binding capacity (TIBC) varies inversely with ferritin level and indirectly measures the amount of blood transferrin (**Figure 3-11**).

- Mean corpuscular volume (MCV) is low in both iron deficiency anemia and anemia of chronic disease; both conditions may have microcytic red blood cells.
- Red Cell Distribution Width (RDW) tends to be increased in iron deficiency anemia but normal in anemia of chronic disease.

Treatment of Anemia of Chronic Disease

- Anemia of chronic disease should be approached with an attempt to correct the underlying cause of the chronic disease.
- Injections of erythropoietin may help when the anemia results from kidney disease where there is underproduction of erythropoietin.
- Blood transfusions are generally not needed because the anemia is usually mild.
- Iron therapy may help as a supplement to erythropoietin, but otherwise does not help because the problem with the disease is that iron is stored but not utilized.

Hemolytic Anemias

Causes of Hemolytic Anemia

Hemolytic anemias are about 5% of known anemias. They can have an *intrinsic* cause, which arises from a defect in the structure of the red blood cell that renders it more fragile, or an *extrinsic* cause, which arises from something that attacks normal red blood cells:

Extrinsic Hemolytic Anemias

Extrinsic causes of hemolytic anemia include:

- **Hypersplenism**, where there is overactivity of the spleen's normal destruction of red cells
- **Mechanical trauma**: RBCs may be damaged passing through *prosthetic heart valves* or through blood clots. *Footstrike trauma* in runners may cause a hemolytic anemia
- Certain **poisons** (e.g. lead, arsenic, copper, snake bite) through their toxic action, rather than an immune effect
- **Autoimmunity,** where the immune system attacks the body's own red cells, mistaking them for foreign substances

Intrinsic Hemolytic Anemias

Intrinsic hemolytic anemias (**Figure 4-2**) arise from a defect within the red blood cell, and are generally hereditary. The kinds of problems include:

- Defective hemoglobin structure (sickle cell, thalassemia, porphyria)
- Defects in the red cell membrane (hereditary elliptocytosis, hereditary spherocytosis)

- A defect in the energy enzyme pathways in the red cell (e.g. G6PD deficiency, pyruvate kinase deficiency)

Sickle Cell Disease

In **sickle cell anemia**, an abnormal globin protein in hemoglobin results from a single DNA nucleotide mutation, leading to an improperly folded globin. This results in sticky sickle-shaped red cells, particularly under conditions of oxygen deprivation. These abnormally sticky cells may clog small blood vessels, depriving tissues of oxygen, depending on where the painful blockage occurs, whether in the small blood vessels of the chest, abdomen, joints, or bones. The condition may also damage the spleen, leaving the patient susceptible to infections. The spleen is in effect a large lymph node and important in fighting infection.

As in many hereditary conditions, homozygotes are far more severely affected. About 10% of American Blacks are heterozygous (*sickle cell trait*) for sickle cell anemia, whereas about 0.4% are homozygous (*sickle cell disease*).

Normal hemoglobin (HbA) has two normal alpha chains and two normal beta protein chains (alpha2, beta2, **Figure 4-3**). In sickle cell anemia, there is a mutation in the beta chain, causing an abnormal hemoglobin (HbS), which has two abnormal beta chains. Individuals with *sickle cell trait* have a mixture of normal HbA and abnormal HbS. Individuals homozygous for sickle cell have HbS but not HbA.

Diagnosis of Sickle Cell Disease

Apart from the usual anemic symptoms, the patient may experience:

- Pain crises, which may manifest as joint pain, chest and abdominal pain that can last hours or weeks
- Inflamed swollen fingers/toes from blocked blood flow
- Pulmonary hypertension/heart disease
- Blood in the urine/renal dysfunction
- Delayed growth or puberty from decreased delivery to tissues of oxygen and nutrients
- Jaundice from hemolysis
- Stroke/seizures
- Retinopathy from blockage of small vessels
- Hepatosplenomegaly
- Splenic sequestration of red cells, with splenic infarction and susceptibility to infections
- Priapism

Laboratory tests show:

- Sickle cells on the blood smear
- Abnormal hemoglobin on hemoglobin electrophoresis
- Serum iron: increased due to excess release of iron from cells during hemolysis
- Ferritin: increased. Hemolysis produces more iron to store
- TIBC: decreased (as expected; TIBC level is always the opposite of ferritin). Increased serum iron (iron overload) results in reflexive reduced transferrin
- RBC count/Hct/Hgb: decreased
- MCV: can be decreased, normal, or increased, taking into account the small size of the sickle cells and the number of reticulocytes, which are relatively large
- RDW: increased. Reticulocytes are prominent in hemolytic anemia and are larger than normal RBCs, generating a wider range of RBC widths
- Abnormal liver function tests (anoxia of liver sinusoids; hepatomegaly)

Treatment of Sickle Cell Disease

- For prophylaxis, the patient should avoid environmental factors that may worsen the condition, particularly dehydration, anoxia from high altitudes, excess exercise, infections, prolonged stasis of the extremities, and skin cooling, which can narrow the blood vessels.
- Sickle cell anemia may be treated acutely with oxygen, pain relievers, fluids, and warmth.
- Folic acid promotes red cell formation.
- Medication to reduce the number of painful crises
  - *Hydroxyurea* (increases the amount of fetal hemoglobin, improving the flexibility of the red cells and reducing sickling and the frequency of crises)
  - *L-glutamine* (may act by reducing oxidative damage in red blood cells, improving their flexibility in traveling through the bloodstream)
  - *Crizanlizumab* (reduces vaso-occlusive events, acting as a monoclonal antibody against *P-selectin*, a cell adhesion molecule)
  - *Narcotic pain relievers*
  - *Voxelotor* (increases oxygen affinity for hemoglobin and reduces polymerization of the abnormal HbS)
  - *Vaccines, antibiotics* in view of susceptibility to infection
- Blood transfusions providing normal red blood cells. Multiple transfusions can lead to iron overload, requiring steps to remove excess iron (*iron chelation therapy*).
- Stem cell transplant with healthy bone marrow
- Gene therapy trial by altering the DNA structure in sickle cell stem cells prior to infusing them back into the patient

Thalassemia

Thalassemia, like sickle cell, is a hereditary disorder of abnormal hemoglobin. In both conditions, there is a microcytic hemolytic anemia.

In *sickle cell anemia*, the mutated hemoglobin, while functional, distorts the cell into a sickle shape, rendering the cell prone to hemolysis and clogging of small blood vessels, resulting in a sickle cell crisis. In *thalassemia*, there is an absence of beta-globin protein in beta-thalassemia and an absence of alpha-globin in alpha-thalassemia. This results in an incomplete hemoglobin, with poorly functioning, fragile cells that are prone to hemolysis.

In both **sickle cell** and **beta-thalassemia** the mutation is to the same gene on *chromosome 11* (**Figure 4-4**), which codes for the beta protein of hemoglobin. In **alpha-thalassemia**, the mutation is in *chromosome 16*, which has two genes for the alpha protein on each chromosome (**Figure 4-4**). This doubling of the alpha-thalassemia gene on chromosome 16, along with mutation variations, allows for the possibility of multiple kinds of thalassemia (over 400), the severity of which depends on how many genes are involved and the mutation variation. There are two main kinds of thalassemia:

Beta-thalassemia: The beta-globin chain is underproduced. It is more common in people of Mediterranean origin.

Alpha-thalassemia: The alpha-globin chain is underproduced. It is more common in people of Asian or African descent.

Both types of thalassemia have major and minor forms, the minor (trait) form being less symptomatic, or even asymptomatic, because it is only a carrier state.

- When only one of the two beta genes is defective, this is **beta-thalassemia minor** (**beta-thalassemia trait**). Symptoms are mild.
- When both beta genes are defective, this is **beta-thalassemia major (Cooley anemia).** Anemia symptoms are moderate to severe and require blood transfusions.

In **alpha-thalassemia**, there are four genes for producing alpha globulin (**Figure 4-4**):

- If only one of the four alpha genes is defective, this is a **carrier state.** The patient is asymptomatic.
- When two alpha genes are defective, this is **alpha-thalassemia trait (alpha-thalassemia minor).** Symptoms are mild. There may be a mild anemia.
- When three of the four alpha genes are defective, the symptoms are moderate to severe (**Hemoglobin H disease, alpha-thalassemia intermedia**), but usually not requiring regular transfusion.
- When all four alpha genes are defective, this is not compatible with life (**hydrops fetalis**).

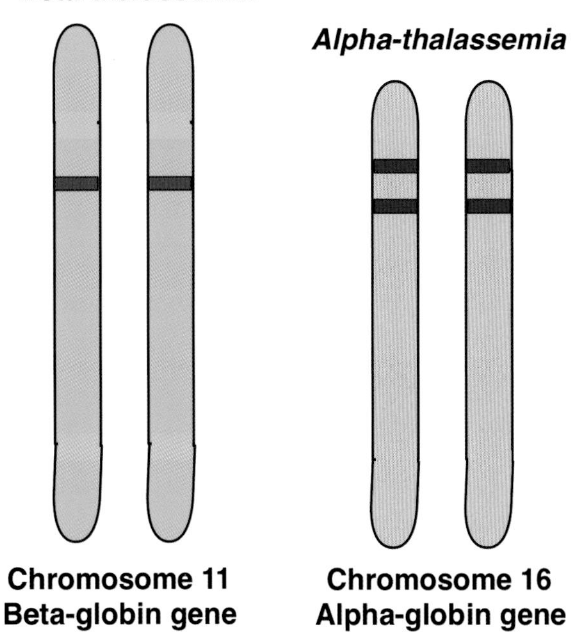

Figure 4-4.

Diagnosis of Thalassemia

Beta-thalassemia major (**Cooley anemia**) is a severe homozygous form of anemia requiring multiple transfusions. The child may appear normal at birth, since fetal hemoglobin, which does not contain beta chains, can temporarily substitute for the abnormal beta chain, but the condition progresses to a severe anemia within a few months after birth. Apart from anemia, beta-thalassemia major is marked by:

- Bone deformities, especially the face, and delayed growth. When the bone marrow has to work overtime to create more and more red blood cells, this "hyperactivity" affects the bones themselves, and can cause them to become distorted, thinner and more easily broken.
- Fatigue
- Yellow skin (jaundice from hemolysis)
- Dark urine (bilirubin)
- Enlarged liver and spleen (*extramedullary hematopoiesis*; also, the spleen enlarges in an effort to clear defective cells)

The *laboratory findings* in thalassemia shows the following:

- While the **ferritin** level is decreased in iron deficiency anemia (there is less need for ferritin in view of the lack of iron), the ferritin level in thalassemia is usually normal because there is no problem with lack of iron. In fact, there may be iron overload from multiple transfusions in thalassemia major, and ferritin may be elevated (**Figure 3-11**)

- While *target cells* can be found in both thalassemia trait and iron deficiency anemia (IDE), basophilic stippling (numerous **basophilic granules** of aggregated RNA, ribosomes, mitochondria that indicates disturbed erythropoiesis) is more characteristic of thalassemia.
- Unlike IDE, the **red cell count** may even be *increased in thalassemia trait* (together with reduced Hct/Hgb) since the red cells continue to be produced (erythropoiesis), but they do not have normal oxygen-carrying hemoglobin, and they are small. The lack of sufficient oxygenation leads to erythropoiesis, but one that produces defective cells. In contrast, in iron deficiency anemia, inadequate dietary intake of iron results in *decreased red cell production* and a *low red cell count*. The red cells are small in both IDE and thalassemia.
- *Bone marrow iron is present in thalassemia trait, but not in iron deficiency anemia.*
- **Treatment with iron** restores hemoglobin values to normal in iron deficiency anemia, but not in thalassemia.
- **RDW** tends to be normal in thalassemia trait but is increased in iron deficiency anemia (**Figure 3-11**).
- **Hemoglobin electrophoresis**. In general, Hgb electrophoresis is not sensitive enough to detect alpha-thalassemia trait, but is abnormal in beta-thalassemia trait.

Treatment of Thalassemia

- Beta-thalassemia minor in most cases is mild and may not require treatment.
- Beta-thalassemia major (Cooley's anemia) may require many blood transfusions, in which case there frequently is iron overload requiring iron chelation to remove excess iron and prevent hemosiderosis (abnormal deposition of iron in tissues).
- Folic acid supplements to facilitate production of healthier red cells, since the hemolysis presents a risk for folate deficiency.
- Splenectomy may help in some people with concomitant large spleen, since a severely enlarged spleen, apart from being uncomfortable and painful in some, can decrease the life span of infused red cells.
- Vaccination and antibiotics are important in view of the risk of post-splenectomy infections.
- Bone marrow stem cell transplantation from a matching donor may be curative.
- Gene therapy trials are in progress. The aim is to substitute a normal beta-globulin gene or increase fetal hemoglobin using stem cells from the patient.
- Endocrine therapy: growth hormone for short stature; gonadotropins for delayed puberty and hypogonadism; insulin if needed for diabetes mellitus, which is common in thalassemia; vitamin D/calcium for osteoporosis.
- *Luspatercept* promotes maturation of red blood cell precursors and lessens the need for as many blood transfusions. Erythropoietin administration may also help.

RBC Energy Pathway Disease

In some anemias, the problem may lie with defects in the red cell enzyme pathways for energy production (**Figure 4-2**). Since red blood cells do not have mitochondria or a Krebs energy cycle to produce energy, their energy is derived from glycolysis (ATP production) and from the HMP shunt (NADPH production). Defects either in glycolysis or in the HMP shunt may result in red cell dysfunction. **Figure 4-5** lists clinical conditions that affect steps in these energy pathways.

Diagnosis and Treatment of RBC Energy Pathway Disease

Diagnosis of the conditions is made by the anemic history and physical exam along with assay of the blood enzyme levels and/or genetic testing.

Treatment may in some cases not be necessary if the condition is mild. In other cases, it may involve dietary change (e.g. avoid fava beans in G6PD deficiency; reduce sucrose, fructose, and sorbital intake in aldolase deficiency); avoid drugs associated with hemolysis. In more severe cases, blood transfusions and splenectomy may be considered.

Red Cell Membrane Disease

A number of gene mutations can affect the primary structure and function of the red cell membrane. These

FIGURE 4-5. HEMOLYTIC DISORDERS OF THE RBC ENERGY PATHWAYS
• G6PD deficiency
• Pyruvate kinase deficiency
• Glutathione synthetase deficiency
• Hexokinase deficiency
• Glucose phosphate isomerase deficiency
• Phosphofructokinase deficiency
• Aldolase deficiency
• Triose-phosphate isomerase deficiency
• Phosphoglycerate kinase deficiency
• Diphosphoglyceromutase deficiency
• Enolase deficiency

decrease the red cell's deformability and render the cells fragile and susceptible to hemolysis. The conditions include:

- Hereditary spherocytosis (spherical RBCs from loss of surface area)
- Hereditary elliptocytosis (ovalocytosis)—cigar-shaped RBCs
- Hereditary stomatocytosis (slit-like notch [stoma] on RBCs)
- Hereditary pyropoikilocytosis (microcytosis, membrane fragmentation)

Treatment of Red Cell Membrane Disease

Treatment may include blood transfusions, folic acid, cholecystectomy (bilirubin gallstones), and total or partial splenectomy.

Hemolytic Porphyria

The **porphyrias** are a group of genetic disorders with different enzyme deficiencies in the pathway that leads to heme synthesis. Every cell needs heme to function. There are *erythropoietic* and *hepatic* forms of porphyria, depending on whether the problem is in the bone marrow or the liver. The marrow synthesizes heme for red blood cells. The liver synthesizes heme as part of the structure of cytochrome P-450 and other enzymes.

Diagnosis of Hemolytic Porphyria

The *hepatic porphyrias* are more marked by neurovisceral symptoms (abdominal pain, neurologic and psychiatric symptoms) and may have abnormal liver function tests. The *erythropoeitic porphyrias* usually are manifest as cutaneous photosensitivity from porphyrin precursors, and hemolytic anemia, which may include splenic enlargement.

The diagnosis is made on the discovery of porphyrin precursors in the urine, which has a dark color, and total porphyrins in the blood. More specific diagnosis can be made by measuring specific tissue enzymes and genetic testing.

As a medical student, I once saw a young woman in the emergency room with a history of frequent episodes of abdominal pain and multiple abdominal surgeries, with no pathology found. She was considered to have a psychiatric problem and had been given phenobarbital. One of the first questions I asked her was whether or not she ever had porphyria. Why ask this? It's a rare disease, affecting only about 5 in 100,000 people and certainly near the bottom of the list in the differential of abdominal pain. However, I had just learned about it in class so it came to mind first. She answered that she didn't have porphyria but had 2 brothers who died of it. Her urine was examined while in the emergency room, and it was positive (dark urine, which darkens further on exposure to light). The chief resident, of course, claimed full credit for the discovery. I learned to include porphyria in the differential when a person has intermittent attacks of nonspecific abdominal pain (especially if there are skin lesions on sun-exposed areas) and may be taking a medication like phenobarbital, which makes porphyria worse. **Figure 4-6** lists other potentially harmful drugs in porphyria.

The *laboratory findings* in hemolytic porphyria show:

- RBC count/Hct/Hgb reduced
- MCV normal
- Serum iron: increased due to excess release of iron from red cells during hemolysis
- Ferritin: increased. Hemolysis produces more iron to store
- TIBC: decreased. Increased serum iron (iron overload) results in reduced transferrin
- RDW: increased. Reticulocytes are prominent in hemolytic anemia and are larger than normal RBCs, generating a wider range of RBC widths
- Abnormal liver function tests with hepatomegaly
- Decreased free haptoglobin. Haptoglobin is a blood protein that binds hemoglobin. Free haptoglobin becomes bound up when there is RBC hemolysis and is low
- LDH increases as it is released from damaged RBCs
- Increased reticulocytes (as occurs in conditions of blood loss, rather than with a problem with RBC production)
- Schistocytes (fragmented red cells), basophilic stippling, and increased circulating nucleated red cells
- Measurement of porphyrins in urine and blood
- Measurement of abnormal enzyme activity in RBCs
- DNA abnormalities for specific kind of porphyria

FIGURE 4-6. POTENTIALLY HARMFUL DRUGS IN PORPHYRIA
Alcohol
Barbiturates
Calcium channel blockers
Carbamazepine
Danazol
Diclofenac
Erythromycin
Metoclopramide
Isoniazid
Phenytoin
Progesterone
Rifampin
Sulfonamides
Valproic acid

Treatment of Hemolytic Porphyria

- Avoid direct sunlight if sensitive. Vitamin D to supplement avoidance of sunlight.
- Avoid triggering factors: alcohol, recreational drugs, fasting, smoking, infections, emotional stress, and medicinal drugs (**Figure 4-6)** that may exacerbate the condition.
- Intravenous heme. Chronic transfusions, though, may result in iron overload, and phlebotomy may then be helpful.
- Hydroxychloroquine and chloroquine, when phlebotomy is not tolerated. They may help by binding to porphyrins and helping remove them.
- Splenectomy in some, where there is significant splenic enlargement.
- Stem cell transplantation from a suitable donor.

Autoimmune Hemolytic Anemias

Autoimmune hemolytic anemias (AIHA) involve an inability of the immune system to distinguish self from non-self. Some cases are secondary to chemicals, medications, infections, some types of cancer; or linked to other autoimmune disease (**Figure 4-7**). Often the cause remains unknown. Some cases may be multifactorial, involving genetic and environmental influences, occasionally showing a tendency to run in families (usually autosomal recessive).

One classification of AIHA focuses on whether they are *warm* or *cold*. In **warm** autoimmune hemolytic anemia (AIHA), the hemolysis occurs at normal body temperature. **IgG autoantibodies** attack the red cells.

In **cold** autoimmune hemolytic anemia, which is less common, cold (32-39°F [0-4°C]) the hemolysis occurs in conjunction with **IgM autoantibodies** (cold agglutinins).

Paroxysmal cold hemoglobinuria is a condition that requires both cold and hot for the hemolysis to occur. The antibody binds to red cells in the cold, but hemolysis does not occur until the body warms up. It may happen with viral infections. This should be distinguished from similar-sounding *paroxysmal nocturnal hemoglobinuria (PNH)*, a bone marrow stem cell mutation, where the red cell surface lacks a protein that normally protects against hemolytic attack by plasma complement. The hemolysis occurs day or night, but the urine is more concentrated in the morning when it appears darker.

Note that cold antibodies are not what are called **cryoglobulins**. Cryoglobulins are antibodies that form

FIGURE 4-7. SECONDARY CAUSES OF AUTOIMMUNE HEMOLYTIC ANEMIA

• **Secondary to other autoimmune diseases** - Rheumatoid arthritis - Systemic lupus erythematosus (SLE) - Sjogren syndrome - Ulcerative colitis - Thyroid disease - Long-term kidney disease - Wiscott-Aldrich syndrome - Chronic lymphocytic leukemia • **Medications** - Penicillins - Cephalosporins - Tetracycline - Erythromycin - Acetaminophen - Ibuprofen - Quinine - Methyldopa - Sulfonamides - Chloroquine	• Infections ***Viral*** - Mycoplasma pneumonia - Epstein-Barr virus (EBV) - Cytomegalovirus - Measles - Mumps - Rubella - Varicella (chickenpox) - Hepatitis - HIV ***Bacterial*** - M. pneumoniae - H. influenza - Legionella - E. coli - Shigella - Bartonella bacilliformis - Clostridium perfringes ***Fungal*** - Aspergillus - Candida - Cryptococcus ***Parasites*** - Malaria - Leishmania - Trypanosomes - Babesia

immune complexes and self-precipitate in the cold. They can cause vascular occlusion under cold conditions as well as a systemic vasculitis, but generally they do not interact with RBCs.

Hemolytic anemias can also be classified depending on whether they are *intravascular* or *extravascular*. *Intravascular* hemolysis occurs in the bloodstream and, when the hemolysis is of the immune sort, it is generally mediated through IgM antibody. IgM antibody has a particular attraction for complement, which in turn bores into the red cell, causing hemolysis directly in the bloodstream. Extravascular hemolysis requires macrophages to destroy the red cells. Macrophages reside outside the bloodstream, where extravascular hemolysis occurs, generally in the spleen and/or liver. Most autoimmune diseases involve extravascular hemolysis, particularly when mediated through IgG, which has a relatively low affinity for complement.

Intravascular hemolysis leaves behind telltale fragments of red blood cells called *schistocytes*. Extravascular hemolysis leaves behind small red cells called *spherocytes* (**Figure 3-18**).

Intravascular hemolysis may occur:

- from trauma of red blood cells passing by prosthetic cardiac valves or passing through blood clots in disseminated intravascular coagulation or thrombotic thrombocytopenic purpura (TTP)
- in G6PD deficiency, a defect in RBC energy metabolism
- in transfusion reactions involved ABO incompatibility
- paroxysmal nocturnal hemoglobinuria where there is a defect in the red cell membrane that allows complement, clustered in groups call MAC (**Figure 6-20**) to hemolyze the cell intravascularly.
- from toxins (e.g. snake bite, clostridial toxins)
- in certain immune drug reactions (e.g. quinine)
- from certain infections (e.g. Plasmodium falciparum)
- in transfusion reactions

In about half of the cases, the cause of the autoimmune anemia is not known.

Diagnosis of Autoimmune Hemolytic Anemia

Apart from usual symptoms of anemia, patients with autoimmune hemolytic anemia may also demonstrate common signs of hemolytic anemia:

- Jaundice (from breakdown of red cells)
- Dark-colored urine (*hemoglobinuria*)
- A feeling of abdominal fullness from an enlarged spleen and liver. Increasing the workload of the liver and spleen to destroy RBCs can lead to splenic and hepatic hypertrophy. Also, a lack of red blood cells can cause the liver and spleen to enlarge to produce their own blood cells (*extramedullary hematopoiesis*).

Laboratory tests in autoimmune hemolytic anemia:

- Serum iron: increased due to excess release of iron from cells during hemolysis
- Ferritin: increased. Hemolysis produces more iron to store.
- TIBC: decreased. TIBC level is generally the opposite of ferritin level. Negative feedback from increased serum iron (iron overload) results in reduced transferrin.
- RBC count/Hct/Hgb: decreased
- MCV (Mean Corpuscular Volume): increased if there are a lot of reticulocytes, which are larger than normal RBCs and contribute to increased MCV. Another reason for increased MCV is the using up of folate in creating more red cells, resulting in a megaloblastic (high MCV) component. In addition, some hemolytic anemias may be accompanied by RBC agglutination, which may be incorrectly interpreted as an increased MCV.
- RDW (Red Cell Distribution Width): increased. Reticulocytes are prominent in hemolytic anemia and are larger than normal RBCs, generating a wider range of RBC widths.
- Abnormal liver function tests with hepatomegaly
- *Low haptoglobin*. Haptoglobin is a blood protein that binds hemoglobin. It becomes bound up during RBC hemolysis when the red cells break open. and haptoglobin is lowered.
- *LDH increases* as it is released from damaged RBCs.
- *Indirect (unconjugated) bilirubin*, a breakdown product of hemoglobin, may be elevated.
- *Increased reticulocytes* from the active production of more red cells
- **Schistocytes** (fragmented red blood cells in intravascular hemolysis) and **spherocytes** (in extravascular hemolysis)
- Pigmented bilirubin gallstones with severe hemolysis
- The **osmotic fragility test** measures the red cells' resistance to hemolysis at different levels of dilution of a saline solution.
- **Hemoglobinuria** and **hemosiderinuria** are more characteristic of intravascular hemolysis than extravascular hemolysis, since the red cells release hemoglobin directly into the bloodstream during intravascular hemolysis.
- Greatly **increased unconjugated serum bilirubin** is more characteristic of extravascular hemolysis, since the unconjugated bilirubin results from

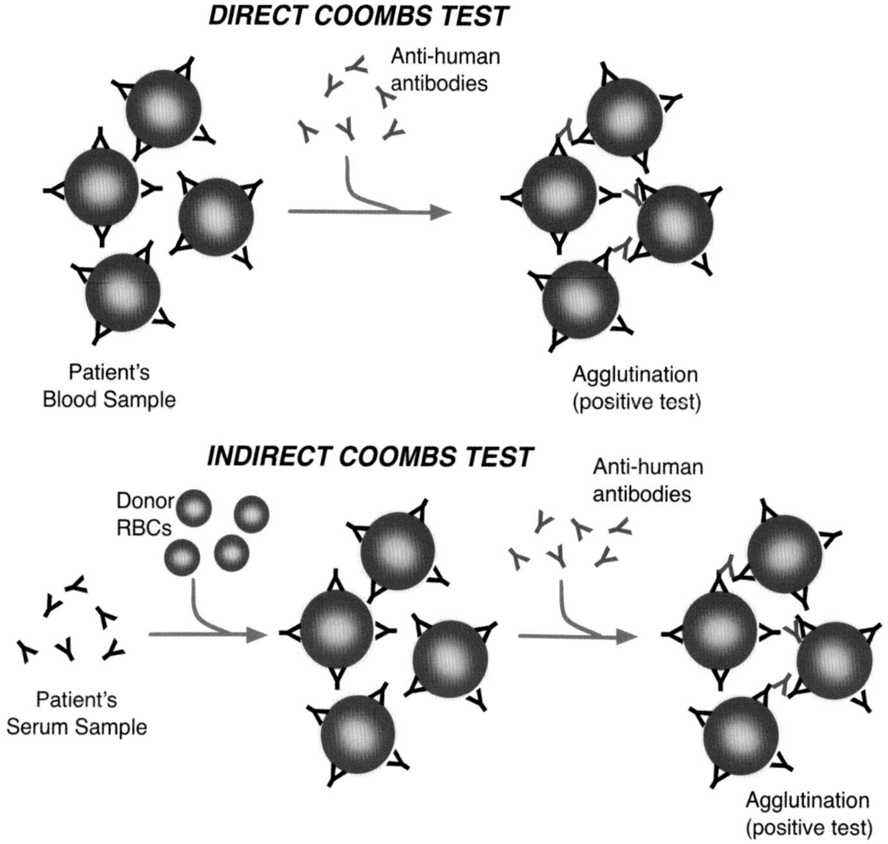

DIRECT COOMBS TEST

Anti-human antibodies

Patient's Blood Sample

Agglutination (positive test)

INDIRECT COOMBS TEST

Donor RBCs

Anti-human antibodies

Patient's Serum Sample

Agglutination (positive test)

Figure 4-8.

breakdown of hemoglobin in macrophages, which are found extravascularly in the spleen and liver. Similarly, splenomegaly (the spleen being extravascular) suggests that the hemolysis involves an extravascular component.

- **Coombs test (Figure 4-8)**: The appearance of reticulocytosis, indirect bilirubin with jaundice, and low haptoglobin, while suggesting hemolytic anemia, does not indicate that the cause is autoimmune. The **Coombs test** detects antibodies against red blood cells.

The *direct Coombs test* (which is the one usually used) detects antibodies that are stuck on the red blood cell surface in autoimmune hemolytic anemia (**Figure 4-8**). The test involves adding anti-human globulin to the patient's *red cells* to see if they agglutinate (a positive test).

The *indirect Coombs test* (**Figure 4-8**) tests for antibodies to red cells that are floating freely in the blood. In this case, the patient's *serum* is incubated with foreign red cells with known antigens on them, followed by the addition of anti-human globulin. If antibodies from the patient's serum combine with the cells and the cells agglutinate, the indirect Coombs test is positive. The indirect Coombs test is used to test if there is a tendency for the patient's serum to agglutinate with blood considered for transfusion (**cross-matching**). The test is also used to determine if a pregnant women is carrying atypical antibodies that might damage the fetal cells (**hemolytic disease of the newborn**).

Treatment of Autoimmune Hemolytic Anemia

- If a drug has caused the hemolysis, avoid the drug.
- Treat other possible causes of the hemolysis: infections, poisons, cold (for cold agglutinin disease), leukemia.
- Treatment of hemolytic anemia may involve immunosuppressives (e.g. corticosteroids, rituximab (a B cell inhibitor), azathioprine, cyclophosphamide, high dose immunoglobulins) if the hemolysis is caused by an immune disease.
- **Eculizumab** (a complement inhibitor) for paroxysmal nocturnal hemoglobinuria reduces RBC destruction. Drawback: It is one of the most expensive drugs in the world (~$500,000/yr).
- Folic acid to assist in forming new cells
- Blood transfusion for severe anemia

- Plasmapheresis to remove antibodies from the blood
- Splenectomy (the spleen destroys red cells) may help if other measures fail to stop the spleen's sequestering and destroying red cells. Treatment by splenectomy is nowadays decreasing given the availability of equally effective second-line medical treatments, e.g. rituximab. Note that splenectomy is usually ineffective in cold agglutinin disease (CAD) since most RBC destruction in CAD does not occur in the spleen.

Sideroblastic Anemia

Unlike sickle cell anemia and thalassemia, which involve defective hemoglobin *protein*, **sideroblastic anemia**, a rare disorder, involves a defect in the incorporation of *iron* into the heme ring of hemoglobin. Instead, the iron accumulates as granules in the red cell mitochondria in a ring-like appearance around the cell nucleus in the bone marrow.

Sideroblastic anemia may be:

- Hereditary
- Acquired as a concomitant of myelodysplastic preleukemic syndrome, particularly in the elderly
- The result of deficiency of copper or vitamin B6 (important in heme synthesis)
- Reversible when due to certain drugs or toxins, e.g. alcohol, lead poisoning, excessive zinc (decreases absorption and increases excretion of copper), isoniazid (interferes with vitamin B6 metabolism), chloramphenicol, or lead poisoning
- Of unknown origin

Diagnosis of Sideroblastic Anemia

- The peripheral blood has red cells with **basophilic stippling** (RNA precipitates, particularly in lead poisoning, one cause of sideroblastic anemia) and **Pappenheimer bodies** (cytoplasmic iron granules) (**Figure 3-18**). The patient, in addition to anemic symptoms, may have an enlarged liver or spleen. There may be bronze-colored skin from iron overload; diabetes mellitus (deposition of iron in the pancreas) and deafness in the hereditary type.
- The red cells have significantly unequal sizes (High RDW) with abnormal shapes.
- The anemia is usually microcytic, but can be macrocytic depending on the specific mutation.
- Serum iron and ferritin are high, while TIBC is low.
- Ring sideroblasts in bone marrow (a ring of iron appears around the nucleus) with a minimum of five siderotic granules around at least one third of the circumference of the erythroblast nucleus).

Treatment of Sideroblastic Anemia

- Rule out reversible causes: alcohol, drug toxicity and other toxin exposure.
- Administer vitamin B6 or folic acid (both important in heme synthesis).
- Blood transfusions, with subsequent chelation or phlebotomy to correct iron overload, as the body has no normal way of removing excess iron
- Stem cell transplant in some

Aplastic Anemia

In **aplastic anemia**, the marrow is wiped out, replaced largely by fat. It can have many causes:

- Heredity (The hereditary form is much less common than the acquired forms and is usually discovered in childhood.)
- Autoimmune disease
- Chemicals/drugs (e.g. benzene, chloramphenicol, phenytoin, quinine)
- Infection (e.g. hepatitis, Epstein-Barr virus, cytomegalovirus, parvovirus B19, HIV)
- Irradiation
- Unknown cause

Diagnosis of Aplastic Anemia

The patient has *pancytopenia*, and, as you would expect, has anemic symptoms from lack of red blood cells, susceptibility to infection from lack of white cells, and a tendency to bleed from lack of platelets.

Fanconi anemia is the most common form of hereditary aplastic anemia, usually diagnosed in children ages 3-14 years. There is a defect in DNA replication and repair. Apart from the aplastic anemia, there are physical abnormalities of the skeleton and kidney, poor growth, short stature, and hypo- or hyperpigmentation of the skin. Other hereditary conditions marked by aplastic anemia include

- *Shwachman-Diamond syndrome*—aplastic anemia, pancreatic dysfunction, skeletal abnormalities, liver impairment, short stature
- *Dyskeratosis congenita*—aplastic anemia, misshapen nails, lacy pigmentation of the upper chest and neck, white patches in the mouth, pulmonary fibrosis, cirrhosis, osteoporosis, and epithelial cancer
- *Diamond-Blackfan anemia*—aplastic anemia, risk for osteosarcoma and other cancers, including acute myeloid leukemia, cataracts, glaucoma, urogenital problems

The *laboratory findings* in aplastic anemia show:

- Pancytopenia
- Low Hgb/Hct
- Bone marrow biopsy shows fewer cells than normal.
- Iron may be elevated due to multiple transfusions, with iron overload that may need chelation therapy.
- Ferritin increased due to iron overload
- TIBC decreased
- MCV: increased (also increased in myelodysplasia)
- Red Cell Distribution Width (RDW): normal
- Reticulocyte count low (cells not being produced)

Treatment of Aplastic Anemia

- If the cause is a drug, remove the drug.
- If the cause of the aplastic anemia is autoimmunity, immunosuppressives are used (e.g. antithymocyte globulin kills T-cells; cyclosporine; corticosteroids)
- Blood transfusions; may need follow-up iron chelation therapy for excess iron
- Bone marrow transplant
- *Eltrombopag*, a drug that induces the proliferation of *platelets* in the bone marrow *Granulocyte colony-stimulation factors* (e.g. filgrastim) may help stimulate *neutrophil* production in the bone marrow.
- *Androgens* stimulate *erythropoietin* production, which may be a reason males tend to have a higher hemoglobin level than females. Androgens sometimes help treat aplastic anemia.
- Administering erythropoetin is not recommended for all kinds of aplastic anemia, as most patients with aplastic anemia already have high erythropoietin levels from the anoxia of the aplastic anemia. Erythropoietin has better use when there is end stage renal disease, where the kidney is not producing sufficient erythropoietin. Erythropoieitin (EPO) is also used for treating marrow suppression from zidovudine-treated HIV-infected patients; patients on marrow-suppressing chemotherapy agents (the EPO decreases the need for transfusions); and for preop in patients at risk of losing blood during surgery.
- Antibiotics for infections due to immunodeficiency
- Immunization with a killed vaccine is safer than live vaccines.

Iron Overload: Hemosiderosis and Hemochromatosis

Hemosiderin is a storage form of iron that follows the breakdown of heme. Like ferritin, hemosiderin stores iron mostly in macrophages. **Hemosiderosis** is an excessive accumulation of hemosiderin, particularly in the lungs and kidneys. The most common cause of hemosiderosis is hemorrhage into an organ and hemolysis. When hemosiderin accumulates in the kidney, it generally does not damage the renal parenchyma, but severe hemosiderinuria may result in iron deficiency. Hemosiderosis is usually not as bad as hemochromatosis.

In **hemochromatosis**, a hereditary disorder, the body absorbs too much iron, which deposits widely (in macrophages and endothelial cells lining the sinusoids of the liver, spleen, and bone marrow). This can damage the liver, heart, spleen, pancreas (diabetes), thyroid (hypothyroidism), gonads (hypogonadism) and skin (bronze discoloration).

Hemochromatosis appears to involve a decrease in the iron-regulatory hormone **hepcidin**, which normally decreases the absorption of iron in the duodenum to protect the body from iron overload. When hepcidin decreases, more iron is absorbed, which, in excess, can cause organ damage.

People with hereditary hemochromatosis can develop anemia because of iron damage to the kidney, anterior pituitary, or bone marrow. The damaged kidney produces less erythropoietin, a hormone needed for red blood cell production (erythropoiesis). An inflamed or damaged anterior pituitary can result in hypothyroidism, which decreases erythropoiesis. And a compromised bone marrow affects red blood cell formation.

Diagnosis of Iron Overload

While many people with hemochromatosis never develop symptoms, others, in their 30's or 40's, will experience symptoms, such as joint and abdominal pain, and weakness. Men are affected more often than women, perhaps because women lose blood in their reproductive years through menstruation, which is one way of losing excess iron.

Complications of iron overload are more likely in hemochromatosis than in the secondary causes of iron overload, such as multiple transfusions and over-consumption of iron. Secondary iron overload can also occur in alcoholic liver disease (low hepcidin production) and hemolytic and other disorders of hemoglobin structure and function (e.g. sickle cell anemia, thalassemia, sideroblastic anemia), which sometimes increase the amount of iron being absorbed. Mild iron overload is often asymptomatic or simply results in fatigue and weakness. Severe iron overload can result in the symptoms of hemochromatosis.

Laboratory findings in hemosiderosis/hemochromatosis include the following:

- The CBC in hemochromatosis will show high serum iron, high serum ferritin, and low TIBC, consistent with the need to store iron and lack of available transferrin binding.
- An MRI may show iron deposits in the liver and spleen, confirmed on liver biopsy.
- Genetic testing can help elucidate the diagnosis.

Treatment of Iron Overload

The body does not have an efficient way to excrete excess iron. Treatment of iron excess by chelation (e.g. with *deferoxamine*), phlebotomy, and avoiding iron supplements are ways to reduce iron.

Polycythemia

Polycythemia is too many red blood cells. You might expect that this shouldn't cause problems, because a little more hemoglobin shouldn't hurt. However, too many red cells can thicken the blood, which decreases blood flow and increases the risk for blood clots, deep vein thrombosis, pulmonary embolus, heart attack, and stroke.

Polycythemia may originate from an overproduction of **erythropoietin**, a hormone that stimulates RBC production. Erythropoietin is secreted by the kidney in response to hypoxia (lack of oxygen) (e.g. living at a high altitude, chronic hypoxia of pulmonary disease, sleep apnea, or cyanotic heart disease). Also, renal cell carcinoma cells can inappropriately produce erythropoietin.

Polycythemia vera, which only rarely runs in families, is an abnormal proliferation of normal red blood cells (and may be accompanied by too many white cells and platelets as well). It is a rare form of blood cancer originating in the bone marrow.

Diagnosis of Polycythemia Vera

The patient develops a pain or full feeling on the left upper abdomen due to splenomegaly from migration of the excess red cells to the spleen. There also is itching, particularly on bathing in warm water, which may have to do with the disease stimulating histamine release from mast cells. The patient may also have bleeding gums (the platelets, while increased, do not function normally). Polycythemia vera may progress to leukemia or myelofibrosis, an uncommon cancerous scarring of the bone marrow.

The *laboratory findings* in polycythemia vera include:

- Elevated Hct/Hgb and RBC count. White cells and platelets may also be increased.
- Normally, erythropoietin (EPO) level, in response to anoxia, increases to stimulate the marrow to produce more red cells. *In polycythemia vera, though, the EPO level is low*, since the problem does not originate in anoxia but in an independent cancerous proliferation of erythrocytes, which reduces the need for EPO.
- Bone marrow aspiration or biopsy confirms the proliferation of too many red cells.
- Gene testing may point to the specific gene associated with the disease.

Treatment of Polycythemia

Polycythemia may be treated by:

- Removal of excess blood (phlebotomy)
- Aspirin to reduce the risk of blood clotting
- Antihistamines or UV light to relieve bothersome itching
- *Hydroxyurea*, which interferes with DNA synthesis, to suppress the bone marrow's ability to produce red blood cells in polycythemia vera. For people not responsive to hydroxyurea, interferon alfa-2b may help by stimulating the immune system to destroy cancer cells. *Bisulfan*, which interferes with cell division, may also be used.

5

Platelets And Blood Clotting

Platelet Function

Platelets are fragments of large megakaryocyte cells in the bone marrow. Despite their small size (about 3 micrometers), platelets perform a number of tasks important to the clotting process. A defect in any of these functions will lead to the risk of bleeding (**Figure 5-1**).

Platelets function as follows in initiating blood clotting with a platelet plug (**Figure 5-1**):

1. Platelets stick to the damaged blood vessel wall, thereby participating in platelet clot formation. **Von Willebrand factor (vWF)**, a glycoprotein, is the initial bridge that connects platelets to the injured vessel wall. **GP Ib-IX** is a protein complex on the platelet cell surface that binds to von Willebrand factor.
   Platelets contract, due to actin and myosin molecules within them, thereby facilitating clot retraction and closure of the damaged blood vessel.
2. Platelets secrete ADP, thromboxane A2, and serotonin, which activate other platelets to become sticky and join in the clot-forming process.
3. Platelet IIb/IIIa protein increases the binding of platelets to fibrinogen and facilitates platelets sticking together.
4. Platelet phospholipids and platelet factor 3 (platelet tissue factor) initiate thrombin formation and are important ingredients in the cascading pathway of blood clotting.
5. Platelets produce fibrin-stabilizing factor (**Factor XIII**), which binds fibrin molecules into a meshwork and strengthens the clot. **Factor XIII deficiency**, a rare autosomal recessive disease, impairs the stability of the clot, which breaks down, with a resulting bleeding tendency.
6. Platelets produce prostaglandins, which have a number of different effects on blood flow and wound healing.

The Blood-Clotting Cascade

When a blood vessel is severed, a series of processes aid in wound closure:

1. Local constriction of the blood vessel, due to local neurogenic causes and local tissue chemical reactions, including platelet release of serotonin
2. Formation of a pure platelet plug, in the case of small wounds to capillaries or very small blood vessels. The pure platelet plug is often sufficient to close small wounds, even without a blood clot.
3. Formation of a blood clot, in the case of larger wounds. A blood clot contains not only platelets, but red blood cells and the end products of the blood clotting processes, particularly fibrin.
4. Wound repair with formation of fibrous tissue

Blood clotting involves a number of clotting factors (**Figure 5-2**):

Factor I (Fibrinogen)
Factor II (Prothrombin)

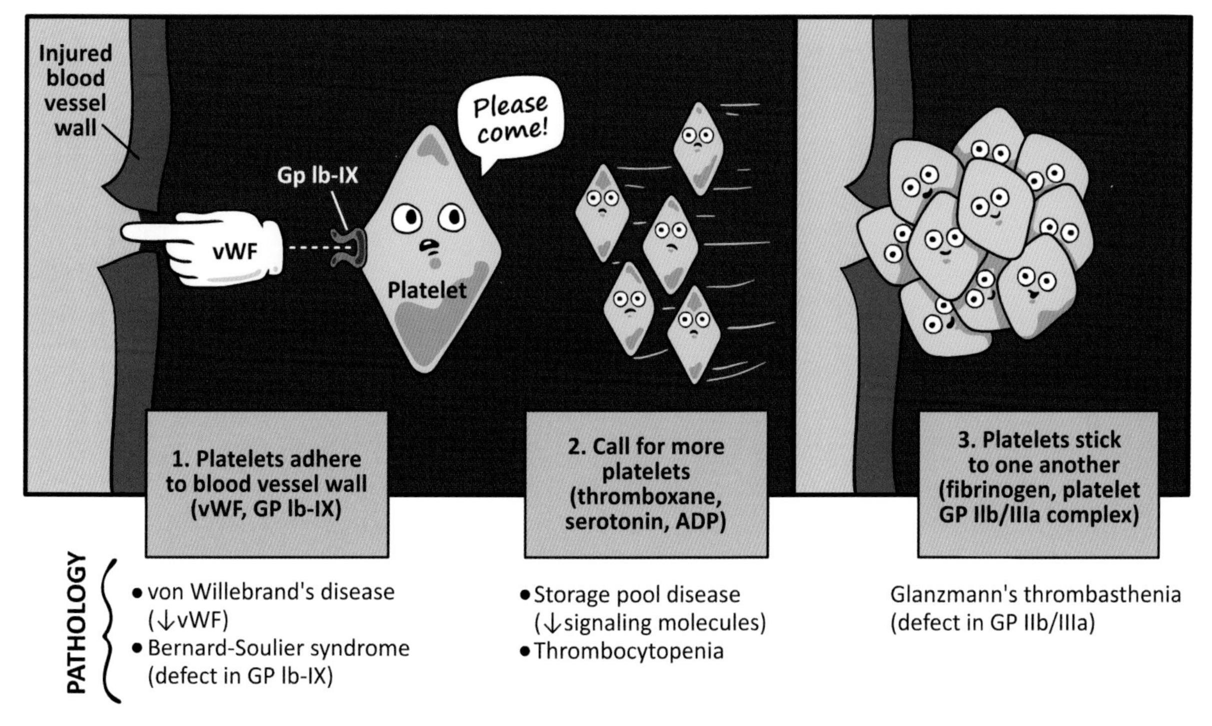

Figure 5-1. *From Berkowitz, A. Clinical Pathophysiology Made Ridiculously Simple, Medmaster.*

Factor III (Tissue thromboplastin)
Factor IV (Calcium)
Factor V (Proaccelerin)
Factor VII (Proconvertin)
Factor VIII (Antihemophilic factor A). Von Willebrand factor is the carrier for Factor VIII.
Factor IX (Antihemophilic factor B)
Factor X (Stuart factor)
Factor XI (Antihemophilic factor C)
Factor XII (Hageman factor)

High molecular weight kininogen (Fitzgerald factor) and **prekallikrein (Fletcher factor)** accelerate the *early* steps in intrinsic path clotting at the level of Factor XII.

Factor XIII (fibrin stabilizing factor) acts at the *end* of the clotting cascade by linking together fibrin molecules and strengthening the clot.

Absence of any factor may lead to the failure of clotting and may be considered a disease in itself.

Blood clotting, apart from the platelet plug, includes a sequence of steps **(Figure 5-2)**:

1. Injury to the blood vessel wall or to the blood cells themselves induces, after a number of initial steps, the formation of **prothrombin activator**.
2. **Prothrombin activator** catalyzes the change of **prothrombin** to **thrombin**.
3. **Thrombin** changes **fibrinogen** to **fibrin** threads, which mix with RBCs, platelets, and plasma to form the blood clot.
4. **Clot retraction**, assisted by platelets, expresses serum. (Serum is plasma minus its clotting factors, such as fibrinogen, etc. which are used up in forming the clot.)

Sounds relatively simple if you start with prothrombin activator. But what are the steps that lead to the formation of prothrombin activator? That is more complex. There an extrinsic pathway and an intrinsic pathway, both of which lead to the formation of the prothrombin activator (**Figure 5-2**).

The **extrinsic pathway** is triggered by tissue **thromboplastin**, which is released from damaged cells in the vascular endothelial wall or outside the blood vessel.

The **intrinsic pathway** is triggered by damage to the blood cells themselves, or by contact of the blood cells with a foreign surface, such as skin or collagen in the wound, or the glass in a blood collection tube. Hence, unless a collected blood specimen is mixed with an anticoagulant, it will clot in the glass tube due to activation of the intrinsic pathway.

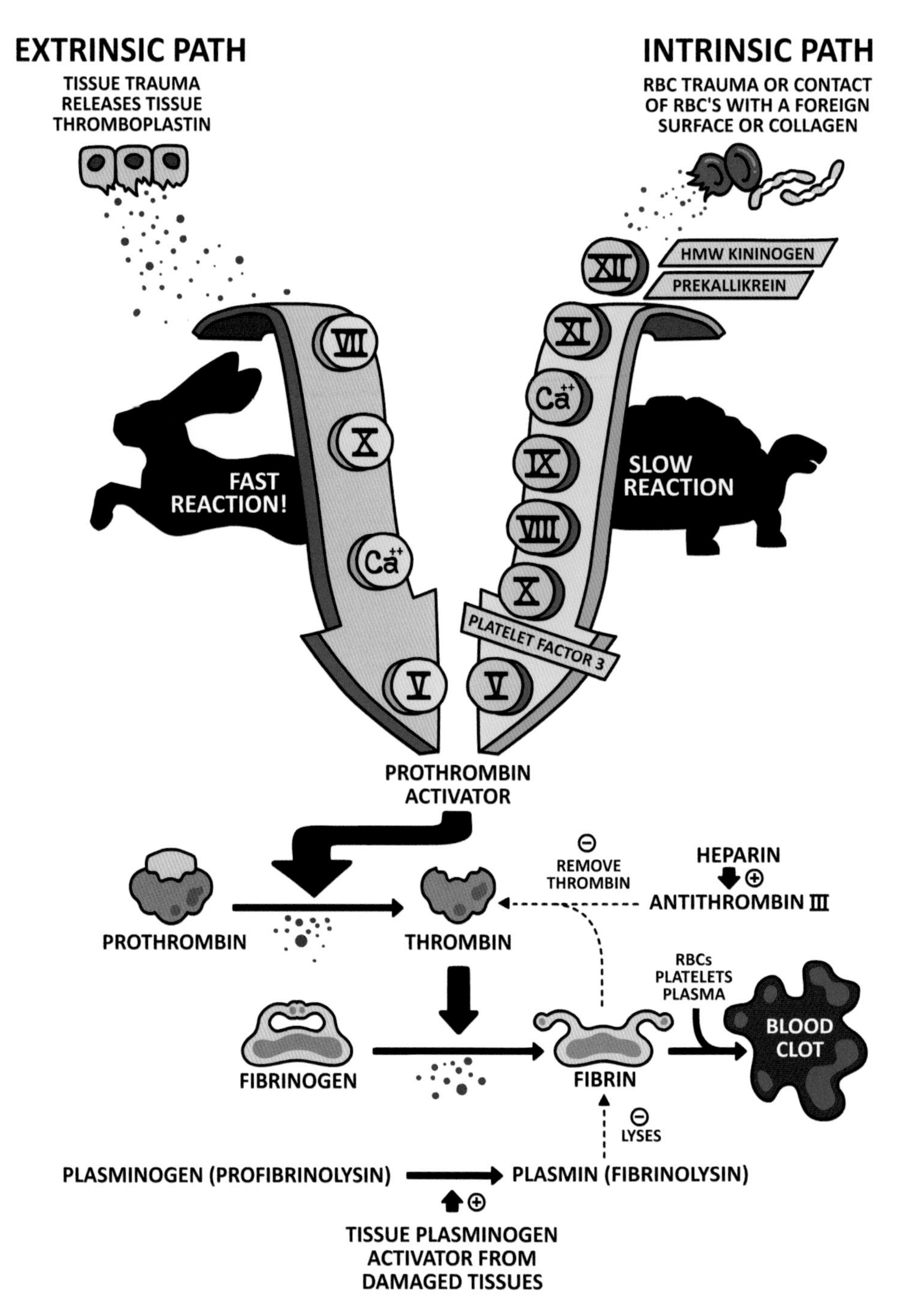

Figure 5-2. Clotting pathways. From Goldberg, S. Clinical Physiology Made Ridiculously Simple, Medmaster.

Both the intrinsic and extrinsic pathways involve a number of clotting factor enzymes that participate in cascading reactions that ultimately lead to the formation of the prothrombin activator.

Figure 5-2 shows a simplified schema of the ingredients of the intrinsic and extrinsic cascading pathways of blood clotting. The extrinsic pathway normally proceeds quickly; blood may clot within 11-15 seconds of activation of the extrinsic pathway. Clotting in vivo may take a number of minutes, however, along the intrinsic pathway, which has more steps than the extrinsic pathway.

Coagulation Tests

Tests for the integrity of the coagulation mechanism are based on whether you want to test the intrinsic or extrinsic pathway, or platelet function, as follows:

1. **Tests of the intrinsic pathway**: The **whole blood clotting time test** measures the time taken for blood to clot in a glass test tube (normally about 2-8 minutes). This tests the intrinsic pathway, which is activated by contact with a foreign surface, such as glass. If any component of the intrinsic pathway is defective, the clotting time will be prolonged over normal controls. You can then identify the particular component that is defective by assaying for particular clotting factors.

More accurate evaluation of the intrinsic and extrinsic pathways may be done using the partial thromboplastin time (PTT) and the prothrombin time (PT):

- The **PTT** evaluates the intrinsic pathway.
- The **PT** evaluates the extrinsic pathway.

The type of tube used to draw blood depends on whether you want to examine whole anticoagulated blood or coagulated blood serum (**Figure 5-3**).

The **partial thromboplastin time (PTT)**, like the whole blood clotting time, tests for the integrity of the intrinsic pathway. Clotting in a test tube is initially prevented by citrating the plasma to remove calcium. The PTT measures the time taken for recalcified citrated plasma to clot in the test tube, usually about 25-35 seconds.

2. **Tests of the extrinsic pathway**: The **one-step prothrombin time (PT)** is the time needed for recalcified citrated plasma to clot in the *presence of tissue thromboplastin*. This "protime" test adds the critical tissue ingredient (tissue thromboplastin) that is necessary to start off the relatively fast extrinsic pathway. The normal protime is about 11–15 seconds, depending on the lab. If the protime is prolonged, in relation to controls, this suggests a problem somewhere in the extrinsic pathway.

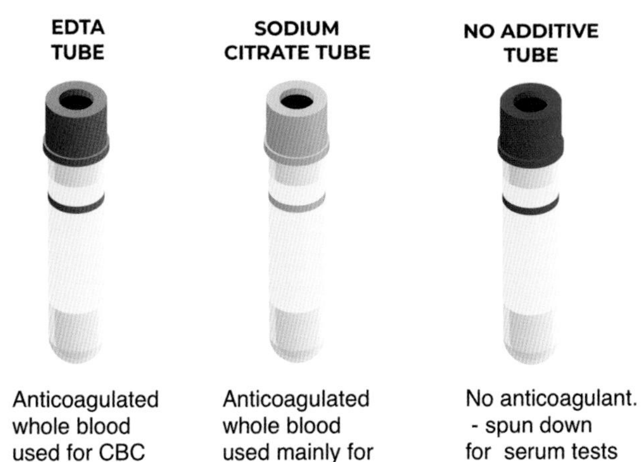

Figure 5-3.

Mnemonic: PTT (the test for the intrinsic path), has an extra "T" intrinsic to it and has more letters than PT (the test for the extrinsic path), corresponding to the intrinsic path having more steps than the extrinsic path.

The **INR** is a way of measuring clotting time that is more accurate than the prothrombin time. While the PT is a relatively simple and inexpensive test, it varies considerably depending on the laboratory and is not a reliable way to determine how well a patient is being anticoagulated on heparin. The INR is more reliable since it is a ratio of the patient's prothrombin time to a series of controls. A normal INR is 1.1 or less, while an INR range of 2.0 to 3.0 is considered an effective range for people taking warfarin for anticoagulation.

Neither the intrinsic nor extrinsic pathway can cause clotting if there is a defect at the end steps of prothrombin-to-thrombin or fibrinogen-to-fibrin. Therefore, *both* the prothrombin time and PTT will be abnormal in

- severe liver disease (the liver manufactures prothrombin, fibrinogen and other clotting factors)
- vitamin K deficiency (vitamin K is necessary for formation of prothrombin and other factors)
- coumarin therapy (coumarin interferes with the formation of the reduced form of vitamin K, and hence the formation of prothrombin and other clotting factors)
- heparin administration (heparin indirectly inactivates thrombin) (**Figure 5-4**).

3. **Tests of platelet function**. Platelet function may be diminished either because of decreased numbers of platelets or because of a deficiency in the functioning of existing platelets.

A platelet count is, of course, useful in assessing the number of platelets. Bleeding may occur with low platelet counts. The **bleeding time** assesses platelet function. A small cut is placed on the patient's forearm, with a blood pressure cuff kept on the arm at 40mm Hg

FIGURE 5-4. PROTHROMBIN (PT) & PARTIAL THROMBOPLASTIN (PTT) TESTS

	PT	PTT
Intrinsic path defect	Normal	Abnormal
Extrinsic path defect	Abnormal	Normal
Liver failure, vitamin K deficiency, coumarin or heparin therapy	Abnormal	Abnormal

to resist venous flow. Normal bleeding time (depending on method) may be 2–8 min. Bleeding time tends to be normal in coagulation disorders of the extrinsic and intrinsic pathway, because the platelet plug operates independently of these pathways and is sufficient in itself to close up such small wounds. If the bleeding time is prolonged, this usually suggests a defect in platelet function.

Clinical Disorders of Clotting

Problems with Platelets

Problems with platelets include:

- **Thrombocytopenia**. The normal platelet count is >150,000 per micro liter. If the count falls below 50,000 (thrombocytopenia), the patient develops numerous widespread dot-like (petechial) hemorrhages and is at great risk for a major life-threatening hemorrhage. The causes include bone marrow depression, immune disorders against platelets, and certain infections.
- **Von Willebrand disease**, the most common of the inherited bleeding disorders, affects about 1 in 100 people globally (but people often don't know about it because symptoms can be mild). It is a hemorrhagic disease with defective adhesion of platelets to the subendothelial collagen in the capillary wall **(Figure 5-1)**. The bleeding time is prolonged. (Aspirin also prolongs the bleeding time as it decreases platelet adhesivity.)

Von Willebrand factor (VWF) acts as a bridging molecule between the platelet and the subendothelium of the blood vessel at sites of vascular injury. VWF is also important in fibrin formation, acting as a carrier for factor VIII in the circulation.

In both **von Willebrand disease**, where there is deficiency in von Willebrand factor, and in **Bernard-Soulier disease**, where there is an abnormality of the GP Ib-IX complex, a protein complex on the surface of platelets that attaches to von Willebrand factor, there is poor platelet adhesion and decreased platelet survival. Von Willebrand disease is usually autosomal dominant, with a normal platelet count. Bernard-Soulier disease is autosomal recessive, with a decreased platelet count. The **von Willebrand factor (vWF) antigen test** measures the amount of vWF, which is decreased in von Willebrand disease.

Increased mean platelet volume (MPV) is found in **immune thrombocytopenic purpura, Bernard-Soulier disease, von Willebrand disease**, and **gray platelet syndrome.** A *large size* may also be a lab artifact due to platelet clumping with EDTA as an anticoagulant.

Decreased mean platelet volume (MPV) is found in the thrombocytopenia of **aplastic anemia** and in **Wiskott-Aldrich syndrome** (an immune deficiency syndrome, with susceptibility to infections and bleeding).

- In **immune thrombocytopenia** (ITP; idiopathic thrombocytopenic purpura) there is an autoimmune destruction of platelets with decreased platelet number, enlarged platelets, a bleeding tendency, prolonged bleeding, and characteristic dot (petechial) hemorrhages in the skin and mucous membranes.
- In **gray platelet syndrome**, an autosomal recessive condition, there is a defect in the platelet granules, along with enlarged and fewer platelets, with a tendency to bruising and prolonged bleeding. **Myelofibrosis** (scarring of the bone marrow) commonly accompanies the condition.
- In **storage pool deficiency**, platelet granules are absent or defective. The platelets then do not release the chemicals needed to attract other platelets, and there is a bleeding tendency. Conditions with platelet storage pool deficiency include:
  - **Chediak-Higashi syndrome** (immunodeficiency, oculocutaneous albinism, platelet dysfunction)
  - **Thrombocytopenic-absent radius syndrome** (absent forearm radius, thrombocytopenia)
  - **Hermansky-Pudiac syndrome** (oculocutaneous albinism, platelet dysfunction)
  - **Wiskott-Aldrich syndrome** (immune deficiency, eczema, thrombocytopenia)
- **Platelet IIb/IIIa protein** is deficient in **Glanzmann thrombasthenia**, a condition marked by bleeding. IIb/IIIa inhibitors are important clot-inhibiting drugs.

Problems with the Clotting Cascade

Interruption of the clotting cascade is an important cause of excessive bleeding:

- **Hemophilia** is a rare bleeding disorder resulting from a defect in the clotting cascade. Factor VIII is missing in the most common hemophilia (Hemophilia A, or classic hemophilia , ~85% of the hemophilias), and Factor IX is missing in most of the remainder (Hemophilia B [Christmas disease], ~15%). Hemophilia C, a much less common form, is a defect in factor XI. Hemophilia A and B are carried as an *X-linked recessive gene* (hemophilia C is autosomal recessive). Therefore, the disease in hemophilia A and B is essentially a male disorder. Males with one hemophilia gene develop hemophilia, whereas females with one hemophilia gene are carriers. Classic hemophilia affects about one in 10,000 males in the U.S.

- **Calcium** (Factor IV) is important to the clotting process, but does not decrease enough to present a clinical clotting problem. However, substances are commonly added to collected blood samples to remove calcium from the blood and prevent the blood from clotting in the tube (e.g. citrate or oxalate ions).
- **Severe liver disease**, such as cirrhosis and hepatitis, can prevent proper clotting, since most of the clotting factors are produced by the liver (e.g. prothrombin, fibrinogen, etc.).
- **Deficiency of vitamin K** may cause bleeding, because vitamin K is important in the formation of prothrombin and factors VII, IX, and X. Normally, we get plenty of vitamin K in the diet, and it is also produced by bacteria in the gut, but in conditions of fat malabsorption (vitamin K is a fat-soluble vitamin), vitamin K deficiency may occur.

Senile purpura is a benign condition found in older people where there is easy bruising and subcutaneous bleeding in the forearm due to fragile skin and blood vessels.

Treatment of Bleeding Disorders

Most bleeding disorders are genetically inherited. There are 3 main approaches in their treatment:

1. Avoid blood-thinning medications.
2. Transfuse platelets or specific clotting factors, depending on whether the problem lies with the platelets or the clotting cascade.
3. Administer medication to improve blood clotting:

- **Desmopressin**. Desmopressin is a hormone that can be administered to help stop bleeding in von Willebrand disease. It causes the release of von Willebrand antigen from the platelets. You can also directly administer recombinant von Willebrand factor. Desmopressin can help treat hemophilia A, because von Willebrand factor carries factor VIII, and desmopressin increases the level of factor VIII, which is deficient in hemophilia A.
- **Recombinant clotting factor VIIa** may be useful in Bernard-Soulier disease.
- **Antifibrinolytic drugs** (e.g. *aminocaproic acid, tranexamic acid*) may also be given orally or intravenously for mild bleeds to retard clot breakdown.
- Prevent platelet destruction when there is an immune component to the disease that destroys platelets.
  - Steroids and immune globulin
  - *Rituximab* is a monoclonal antibody that acts on the CD20 antigen on B lymphocytes. It is used as a treatment in a variety of lymphoproliferative disorders as an immunosuppressive to decrease immune reactions, including those that destroy platelets.
  - Drugs to increase platelet production (e.g. *romiplostim, eltrombopag*)
  - Platelet transfusion
  - Splenectomy in some patients, to eliminate platelet destruction by the spleen

Excessive Clotting

Can clotting be excessive? The body *normally* has a number of negative feedback mechanisms that prevent excessive clotting:

1. Fibrin in the clot absorbs excess thrombin.
2. A globulin in the clot area called **anti-thrombin III** (**Figure 5-2**) inactivates excess thrombin and forms a check on excessive thrombosis.
3. **Heparin** is a powerful anticoagulant produced by mast cells and basophils. Heparin enhances the activity of antithrombin III.
4. **Plasminogen** (**profibrinolysin**) is a plasma protein that, when activated, becomes **plasmin** (**fibrinolysin**). Fibrinolysin *lyses* fibrin (**Figure 5-2**) and helps remove the clot. The process of plasminogen activation is relatively slow, occurring at least a day after the clot has formed to allow time for the initiation of wound healing. It is activated by **tissue plasminogen activator**, which is released by the damaged tissues.

Excessive clotting (thrombophilia) may occur in:

- **Disseminated intravascular coagulation** (**DIC**). In this condition, small clots develop throughout the circulation and block small vessels. This uses up platelets and clotting factors, and the patient is then susceptible to bleeding even in the face of the excess clotting. What should the physician do – try to decrease the clotting or try to decrease the bleeding? A number of things can cause DIC: infection (COVID-19 virus is one; vaccination against COVID-19 is being studied as a possible rare cause), recent surgery or anesthesia, blood transfusion reaction, some cancers (e.g. certain leukemias), pancreatitis, liver disease, and childbirth complications. The labs show a low platelet count, decreased fibrinogen, and increased *plasma D-dimer* (a protein fragment from clot breakdown).
  DIC is treated by correcting its underlying cause, platelet replacement, coagulation factors (in fresh plasma), fibrinogen to control the bleeding, and by cautious use of heparin to control the clotting

in patients at risk for venous embolism. Steroids are an option for dealing with an inflammatory component.

- **Antiphospholipid antibody syndrome**, in which there is an autoimmune overactivation of blood clotting factors. It is a cause of recurrent abortion from placental infarction. It is treated with anticoagulation.
- Deficiency of one of the natural anticoagulants that normally temper the clotting cascade (antithrombin, protein C, protein S, protein Z). It is treated with anticoagulation (e.g. warfarin).
- Mutation in the prothrombin gene that results in excess production of prothrombin. It is treated with anticoagulation (e.g. warfarin, heparin).
- **Dysfibrinogenemia**. Different mutations result in an abnormally functioning fibrinogen. It may manifest as excess clotting or excess bleeding, depending on the mutation. Treatment may include heparin for a clotting problem, or fresh frozen plasma or plasma cryoprecipitate transfusion to provide clotting factors to help control bleeding.
- Conditions associated with sluggish blood flow, e.g. sickle cell anemia, myeloproliferative disorders, where there are too many cells in the blood (e.g. polycythemia vera and essential thrombocythemia). (See respective chapters for treatment approaches.)
- Cancer, which sometimes can activate the coagulation system
- Pregnancy, which may incur a physiologic mechanism to guard against postpartum hemorrhage
- Estrogen use
- Obesity
- Simply not moving around enough, causing blood to pool in the legs, as in prolonged bed rest or sitting, or conditions that increase back pressure on the veins, as obesity and pregnancy. Anticoagulation may be indicated for clotting.

Treatment of Excessive Clotting

Anticoagulant drugs that interfere with *platelet aggregation* include:

- **Aspirin**, which interferes with blood clotting by inhibiting cyclooxygenase. Cyclooxygenase is an enzyme that is important in the synthesis of thromboxane A2, which activates platelets to become more sticky in the clotting process
- **Clopidogrel** (*Plavix*) inhibits the binding of ADP to its receptor on platelets, and this inhibits platelet aggregation.
- **Glycoprotein (GP) IIb/IIIa receptor inhibitors** (e.g. *abciximab, eptifibatide, tirofiban*) prevent the binding of activated platelet GP IIb/IIIa receptors to fibrinogen and von Willebrand Factor and prevent platelets from sticking together.

Anticoagulant drugs that interfere with the *clotting pathways* include:

- **Heparin** has immediate anticoagulant effects by enhancing the activity of antithrombin III, and is administered intravenously.
- **Coumadin** (**warfarin**) acts by a different mechanism. It competes with vitamin K and inhibits the production in the liver of prothrombin and other clotting factors. It is given orally, but requires one or more days to take effect, since you must wait for the existing prothrombin to be depleted.
- **Bivalirudin** is an example of a **direct thrombin inhibitor**, which binds to and inhibits thrombin (thereby preventing fibrinogen from forming fibrin), reducing clot formation.
- **Plasminogen activators** (such as *streptokinase, alteplase, reteplase,* and *tenecteplase)* may be used to dissolve coronary artery clots in patients with acute myocardial infarction.

6

White Blood Cells and the Immune System

White blood cells play a key role in immunity. The field of Hematology overlaps Immunology, and both fields need to be considered to properly understand the role of white blood cells. The following is a review of the major features of the immune system.

The main function of the immune system is protection from foreign organisms. Its key strategy is to distinguish self from non-self, and to eliminate the foreign. An **antigen** is a foreign substance that can induce an immune response.

Although there are highly specific defenses, such as antibodies, much defense occurs through less specific mechanisms that arose prior to the time that antibodies evolved. These less specific mechanisms still persist today as vital components of the specific immune response. These less specific mechanisms are called **natural immunity**, as opposed to the more specific **adaptive immunity** of antibodies and other actors in the immune response.

Natural (innate) Immunity

Natural (innate) immunity from microorganisms, apart from obvious mechanisms such as stomach acidity and the physical barrier of the skin, include a diversity of molecules and cell types:

FIGURE 6-1. ACTORS IN THE IMMUNE RESPONSE

Natural Immunity		Adaptive Immunity	
Molecules	**Cells**	**Molecules**	**Cells**
Interferon Lysozyme Complement C-reactive protein Prostaglandins Kinins Leukotrienes and other cytokines	Macrophages Microglia Dendritic cells Langerhans cells Kupffer cells Alveolar macrophages Neutrophils Eosinophils Basophils Mast cells Platelets Natural killer cells Vascular endothelial cells Kidney mesangial cells Reticular cells	Antibodies	B lymphocytes T lymphocytes: - T helper cells - T cytotoxic cells - T suppressor cells

Molecules of Natural Immunity

Interferons: a group of glycoproteins that, among other things, kills viruses and in general activates macrophages to do a better job in killing phagocytosed microorganisms. Most cells can secrete interferon, an important cellular defense mechanism, considering that viruses must live within cells.

Lysozyme: a natural antibiotic against bacteria. It is produced by macrophages and neutrophils, and attacks the bacterial cell wall.

Complement: includes at least 15 proteins, mostly enzyme precursors, which are produced in part in the liver, found in the serum, and, when activated, undergo cascading chain reaction conversions that are important in the immune response (**Figure 6-19**).

Figure 6-2. Functions of complement. The Complement Can is spilling "OIL" (Opsonization, Inflammation, Lysis):

- **Opsonization** (complement C3b) is a "gluing" function of complement that attaches bacteria (or red blood cells in the case of certain hemolytic anemias) to neutrophils and macrophages, which then can more readily eat (phagocytose) them. (**Mnemonic:** C, 3, and b "stick" together as all the letters rhyme.) Opsonization is particularly important to attach the tough-to-attack encapsulated bacteria to phagocytic cells. Complement activation can also activate natural killer cells to attack and kill the cell by releasing cytotoxic granules, rather than killing by phagocytosis.

Figure 6-2. *Complement functions. From Goldberg, S. Clinical Physiology Made Ridiculously Simple, Medmaster.*

Although complement does not act with the specificity of antibodies, it does indirectly participate in the specific elimination of antibody/antigen complexes by attaching such complexes to phagocytic cells, which then can eliminate them more efficiently.

- **Inflammation**. Other forms of complement can stimulate an acute inflammatory response, inducing histamine release from mast cells and basophils. This renders blood vessels more leaky so that neutrophils and other inflammatory cells, as well as complement protein, can enter the damaged tissue. Certain complement components may in themselves exert a chemotactic (chemical attraction) influence on neutrophils.

  The general inflammatory response is complex, including as its goal not only the defeat of invading organisms, but wound healing, too. Scavenger cells, such as macrophages and neutrophils, not only take care of invading organisms, but also clean up damaged tissue. Intertwined with the antibiotic effect of the immune system are clotting mechanisms, which stop bleeding, with the formation of fibrin (see **Chapter 5**), into which fibroblasts migrate and produce collagen, which helps form a healing scar.

- **Lysis**. A form of activated complement, the **Membrane Attack Complex (MAC) (Figure 6-19)** can bore its way through the cell walls of bacteria (or red blood cells in autoimmune hemolytic anemia) and destroy them through cell leakage (lysis).

Figure 6-3. The inflammatory response. Symptoms are *calor* (heat), *rubor* (redness), *tumor* (swelling), *dolor* (pain) and *functio laesa* (loss of function).

C-reactive protein: a type of serum globulin that is produced in the liver and increases during acute tissue injury or inflammation. It binds to the surface of bacteria and facilitates phagocytosis. It activates complement. The acute phase of inflammation can often be detected in the laboratory through measurement of C-reactive protein.

Prostaglandins and **leukotrienes**: fatty acids that may be released by damaged cells, including mast cells. They have a variety of functions, such as enhancement of the inflammatory response by increasing vascular permeability, chemotaxis (chemical attraction) of neutrophils toward the wound site, and stimulating pain endings.

Kinins: serum proteins that stimulate vascular dilation and permeability during inflammatory reactions.

Interleukins: Among other things, interleukins

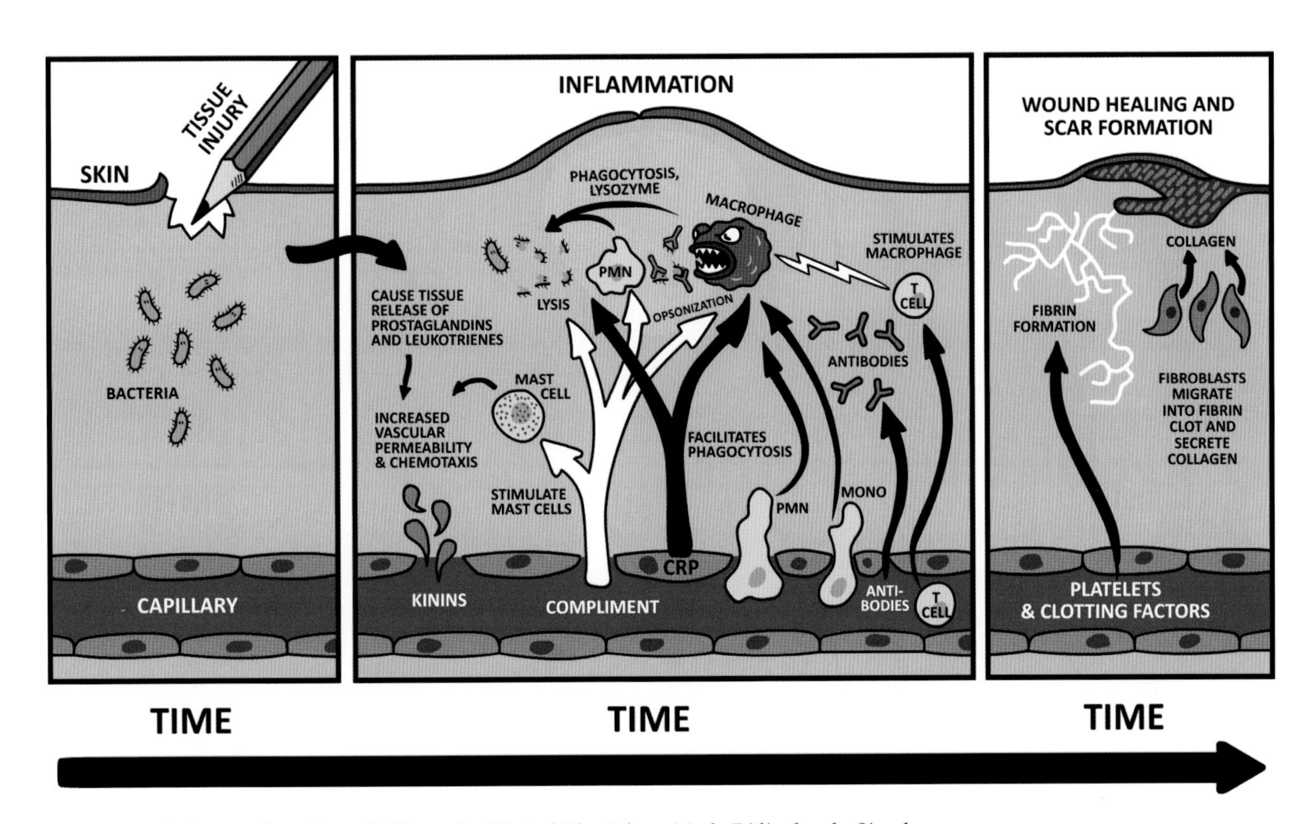

Figure 6-3. Inflammation. From Goldberg, S. Clinical Physiology Made Ridiculously Simple, Medmaster.

stimulate the proliferation and maturation of lymphocytes during the immune response.

Colony-stimulating factors (CSFs): Among other things, CSFs stimulate the proliferation and development of granulocytes, monocytes and macrophages. **Erythropoietin** is a CSF that is produced in the kidney and stimulates red cell production.

Tumor Necrosis Factors (TNF): Among other things, TNF can help stimulate acute inflammation.

TGF-beta: Among other things, TGF-beta can help stimulate wound healing.

Cytokine is a general term, usually signifying a protein hormone that affects the function of cells lying near the cytokine's cell of origin. A number of the above-mentioned molecules are cytokines, including interleukins, interferons, colony-stimulating factors, tumor necrosis factors, and TGF-beta. The many kinds of cytokines often overlap in their functions.

Cells of Natural Immunity

Figure 2-1 reviews cell lineage in the immune system.

Monocytes, which are found in the blood, are the precursors of tissue macrophages, which participate in natural immunity. The various kinds of tissue phagocytic cells include:

1. **Macrophages**. Macrophages, without the assistance of antibodies, can often recognize and directly phagocytose many kinds of bacteria, in addition to recognizing and cleaning up damaged tissues. Macrophages can also kill many bacteria without phagocytosing them, by secreting toxic enzymes, e.g. *lysozyme*. The Golgi apparatus inside macrophages spins off lysosomes, which contain lysozyme and other digestive enzymes that are either secreted or used internally to digest bacteria. For internal digestion, the lysosome fuses with a *phagosome* (a vesicle containing a phagocytosed microorganism) to form a *phagolysosome*, where the microorganism is killed and digested internally (**Figure 6-4**). Macrophages in the liver and spleen normally identify and phagocytize red blood cells that reach their life span of about 120 days old. Macrophages in the liver and spleen also phagocytize bacteria or red cells with attached antibodies or antibody-complement complexes in autoimmune hemolytic diseases (**Figure 6-13**).

   Figure 6-4. Action of lysozyme either externally or internally to kill bacteria.
   Figure 6-5. A **granuloma**. In chronic inflammation, macrophages may move together to form a granuloma, another means of defense. Granulomas contain *epithelioid cells* (macrophages that have acquired a large amount of cytoplasm) and *multinucleated giant cells* (a fusion of a number of macrophage cells). Fibrosis with calcification frequently occurs around granulomas. Granulomas wall off infectious material from the rest of the body. They may form if the phagocytic cells cannot digest the foreign material (e.g. asbestos, silica). Indigestible immune complexes may also induce granulomas. Granulomas also arise when there is a continuing irritant stimulus. Thus, granulomas

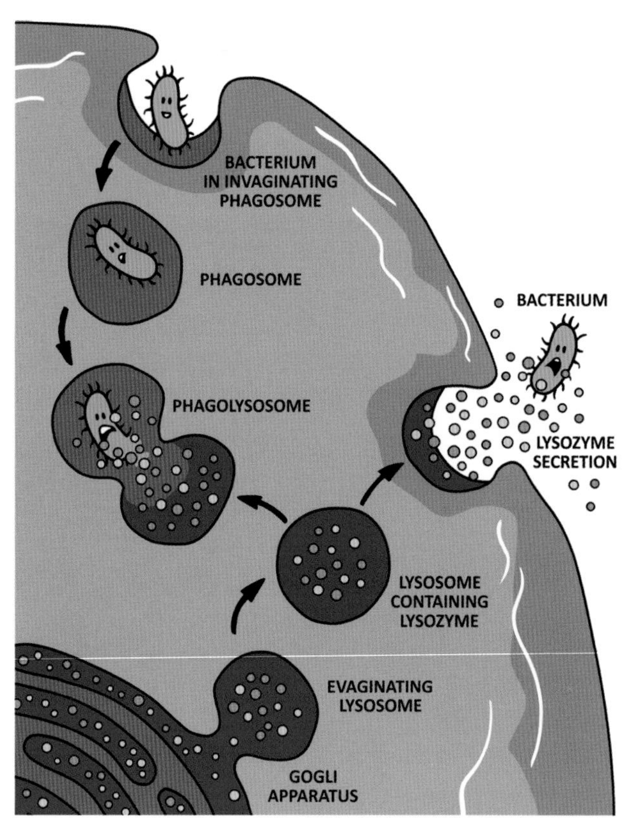

Figure 6-4. Lysosome action on bacteria. From Goldberg, S. Clinical Physiology Made Ridiculously Simple, Medmaster.

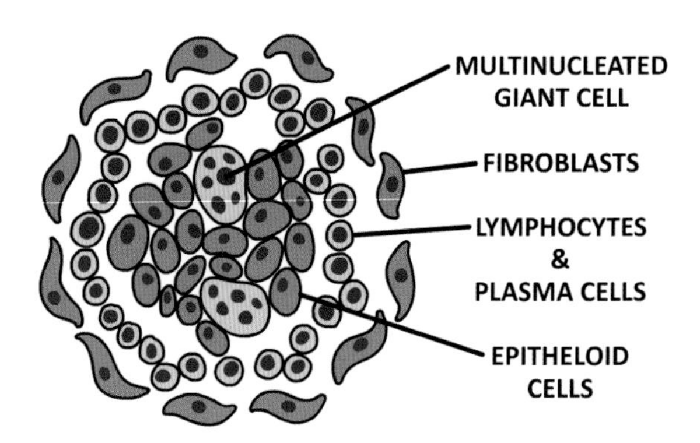

Figure 6-5. Granuloma. From Goldberg, S. Clinical Physiology Made Ridiculously Simple, Medmaster.

may form in certain autoimmune diseases, where there is a persistent reaction to the host's antigens.

2. **Microglia** are phagocytic cells found in the brain.
3. **Dendritic cells**, found in the mucosa, skin, spleen and lymph nodes, besides being phagocytic, play an important role in presenting antigen to T cell lymphocytes (discussed later).
4. **Langerhans cells**, found in the skin, are precursors of dendritic cells. They migrate to draining lymph nodes from areas of acute inflammation, carrying with them antigens for presentation to lymphocytes.
5. **Kupffer cells** are phagocytic cells within blood sinus walls in the liver.
6. **Alveolar macrophages** are found in the lung.

T helper lymphocytes (discussed later) are sophisticated cells that need to be properly introduced to antigen in order for them to react to the antigen; the antigen generally needs to be presented to the T helper cell by another cell. A number of phagocytic cells (particularly macrophages, dendritic cells, and Langerhans cells) not only are phagocytic but also present antigens to T lymphocytes. (Endothelial cells and B lymphocytes can also present antigens to T cells.)

Note that stem cells in the bone marrow give rise to monocytes, which in turn give rise to the above-mentioned categories of phagocytic cells. Bone marrow cells also give rise to *granulocytes* (neutrophils, eosinophils, and basophils, which contain visible *granules*) and platelets.

Neutrophils are scavenger white cells. These short-lived cells (living only up to 1–2 days) quickly exit the bloodstream into inflamed tissues. (Lymphocytes and monocytes arrive last.) Neutrophils make up about 40-60% of the blood WBCs. Neutrophil granules contain potent bactericidal enzymes used for internal digestion of bacteria.

Eosinophils increase in number in parasitic (notably worm) infestation. They contain cell surface receptors for immunoglobulin E (IgE), which is produced at high levels in parasitic infection. In response to antigens that attach to eosinophil cell-bound IgE, eosinophilic cell granules release substances that are toxic to parasitic worms, which often are resistant to neutrophil and macrophage lysozymes. Eosinophils also increase in number during allergic reactions in which there is stimulation of IgE production. Like neutrophils, eosinophils can exit through blood vessel walls into the tissues, where their cytoplasm releases cytotoxic granules.

Basophils are granulated white cells in the blood that also can exit the bloodstream into the tissues. Their granules contain heparin and vasoactive amines, such as histamine and serotonin, which increase vascular permeability and allow inflammatory cells and complement to enter the tissues from the bloodstream. They are not phagocytes. Serotonin is a chemoattractant that recruits immune cells to the site of inflammation.

Mast cells, unlike the circulating basophils, are not found in the blood. They are tissue-based, in connective and mucosal tissues, and are believed to mature in those tissues from immature precursor cells from the bone marrow. Like basophils, their granules contain heparin and powerful inflammatory substances (e.g. histamine, serotonin) which, when secreted, increase vascular permeability and allow inflammatory cells and complement to enter the tissues from the bloodstream. Both basophils and mast cells have surface IgE receptors. Secretion occurs when antigen binds to cell-bound IgE on basophils and mast cells. (**Figure 6-15**). Mast cells also function as phagocytes and as antigen-presenting cells (**Figure 6-9**).

Platelets, although usually considered in the context of blood clotting, can also phagocytose antigen/antibody complexes.

Natural killer cells are lymphocyte-like cells that can kill virus-containing cells, but do so without the high-specificity characteristics of the B and T lymphocytes.

There are other kinds of phagocytic cells that arise not from bone marrow stem cells, but from the tissue mesenchyme throughout the body. These include:

Vascular endothelial cells, which can produce and respond to cytokines, and, under certain circumstances, present antigen to lymphocytes.

Bone osteoclast cells, which resorb bone.

Kidney mesangium cells in the renal glomeruli, which can phagocytose antigen/antibody complexes that deposit there.

Reticular cells of the lymph nodes, spleen, thymus, and bone marrow.

Adaptive (acquired) Immunity

As higher animals evolved, invading organisms also evolved and developed defense mechanisms against the simple natural immunity of the human host. Faced with the difficulty of defending against the rapid mutations of bacteria, viruses, and other organisms, the more advanced organisms adopted the strategy of developing specific antibodies against all possible antigens.

Antibodies are proteins produced by plasma cells, which are derived from B lymphocytes. Antibody production constitutes *humoral immunity*. A second form of adaptive immunity, *cell-mediated immunity*, requires direct contact of the antigen with lymphocytes, notably T lymphocytes, and is not mediated through antibodies.

Humoral (antibody-mediated) Immunity: B Lymphocytes

Figure 6-6. Structure of the IgG antibody molecule. The myriad specificities of the antibodies are determined by the variable portions of the antibody molecule, which reside in the Fab portions of the molecule. There are two Fab sections—the two arms of the "Y," each of which contains a heavy and a light chain. The variable portion (V) is at the free end of each Fab segment arm. The rest of the molecule is constant (C) in all the IgG molecules of the individual.

An antigen must be a macromolecule, either a large protein or polysaccharide, in order to activate lymphocytes to generate antibody formation. If an antigen is too small to generate an immune response by itself, it is called a *hapten*. If it is large enough to generate an immune response, it is called an *immunogen*.

Figure 6-7. The structure and function of the major antibody groups, or **isotypes**. A **mnemonic** for the various isotypes is "GAMED"—immunoglobulins G, A, M, E, and D, or IgG, IgA, IgM, IgE, and IgD. The particular isotype is determined by the constant, Fc portion.

- **IgG** is the main, classic immunoglobulin.
- **IgA** is the main antibody in secretions and plays a significant role in first-line defense at the mucosal level.
- **IgM** is the main antibody in the initial "primary" immune response and shows good complement activation in view of its large ("Magnum") size, consisting of five IgG-like subunits. Its multiple arms enable it to better attach to and produce agglutination of surrounding red blood cells in disease where RBC agglutination occurs.
- **IgE** is found in allergy and worm infestation. Its Fc region binds to eosinophils, basophils and mast cells. It is a significant mediator of allergic hypersensitivity reactions when antigen binds to surface-bound IgE.
- **IgD** is an antigen receptor mainly found on the surface of B lymphocytes. It functions as an antigen receptor, signaling the B cells to activate and participate in body defense.

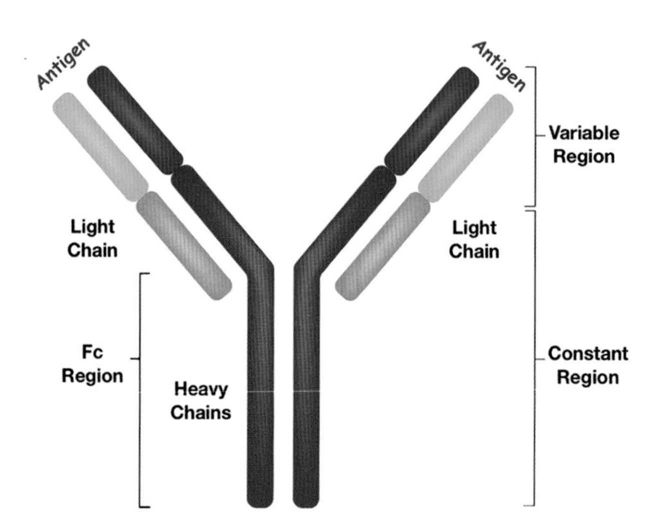

Figure 6-6. Antibody structure. From Goldberg, S. Clinical Physiology Made Ridiculously Simple, Medmaster.

It was no small evolutionary task to develop millions of different antibodies against all possible antigens, considering the limited number of genes (about 20,000-25,000) in the human genome. According to the *clonal selection theory*, the human immune system, prior to any antigen stimulation, already contains numerous clones of antibody-producing cells, each different clone confined to producing only one specific antibody for one specific antigen. The immune system, thus, prior to being introduced to foreign antigens, already contains the antibody repertoire against all of them. In addition to the enormous amount of information necessary to create such a repertoire, such a strategy had to result in antibodies against only foreign antigens, while leaving the host cell antigens intact.

The body resolved the problem of this enormous information requirement in much the same way as one obtains card combinations in a shuffled deck. A single deck contains 52 cards. If each card represented a single gene for a single antibody, one could obtain only 52 antibodies. However, if an antibody consists of a combination of one club, one spade, one heart, and one diamond, each suit reflecting a different gene for a segment of the antibody molecule, then the number of combinations is vastly increased. The variable portions of the heavy and light chains of the antibody molecule are each formed by such a mechanism: the subdivisions of each variable portion are formed independently and shuffled to produce many different variable segments. The combination of a variable portion of a heavy chain with a variable portion of a light chain compounds the possibilities. Mutations can also increase the variations. In addition, there are the five antibody isotypes (GAMED), which are determined by the constant portion of the antibody molecule. Thus the number of kinds of antibody molecules is compounded by the association of the

Immunoglobulin	Structure	Function
IgG	**Y monomer**	Prominent in secondary response
IgA	**dimer**	Prominent in secretions
IgM	**pentamer**	Prominent in primary antibody response
IgE	**Y long Fc**	Prominent in worm infestations & allergies
IgD	**Y monomer**	Receptor on B-lymphocyte

Figure 6-7. *Antibody groups.*

myriad variable portions with any of the five basic constant portions.

In the end, only one specific antibody is produced by each antibody-producing B lymphocyte. **Monoclonal antibodies** are identical antibody molecules synthesized by a clone of cells that arose from a single antibody-producing cell. Monoclonal antibodies may be created in tissue culture conditions, but may also be found in certain lymphocytic tumors. Monoclonal antibody formation is not, however, the typical response in vivo to a single antigenic stimulus. What usually occurs is that a single antigen stimulates a variety of closely related lymphocytes to produce different antibodies that more or less resemble the correct fit for the antigen. (Each of those antibodies may attack different antigens that more or less resemble one another.) The subsequent rise in plasma antibody in a typical electrophoresis is then broader than a single sharp spike.

It is important for the immune mechanism to distinguish self from non-self to avoid adverse interaction with the individual's own cells. One mechanism of such self-tolerance of B or T cells is the elimination, or inactivation, of those B and T cells that interact with self-antigens. Such elimination is believed to occur during fetal development, but also in later life as immature lymphocytes arise. It appears that a given immature B or T cell interacts with its corresponding self-antigen, leading to the death, or at least inactivation, of the self-reactive lymphocyte. The exact mechanism is still unclear, but it is believed that self-reactive T lymphocytes are eliminated in the thymus gland, where T cells mature under the influence of thymic hormones. In autoimmune diseases, there is a defect in the ability to distinguish self from non-self for certain kinds of cells.

The key cell responsible for adaptive immunity is the **lymphocyte**, each of which specifically can act against only one particular antigen. The specific lymphocytes include B cells and T cells:

B cell lymphocytes originate in the **B**one marrow. They **B**low out free-floating anti**B**odies into the **B**loodstream. (Actually, it is the plasma cell, derived from the B cell, that produces the antibody.) When an antigen contacts a matched B cell, it stimulates the specific surface immunoglobulin on the B cell, which in turn stimulates the B cell to divide and form plasma cells and B memory cells.

Plasma cells produce the specific antibody against the antigen. Contrary to the name, few plasma cells are found in the plasma. Most reside in lymphatic tissue, particularly the lymph nodes and the spleen. The secreted antibody enters the bloodstream, though. *Humoral immunity*, as mentioned, refers to immunity based on B cell generation of circulating antibody, in contrast to *cell-mediated immunity*, which requires direct contact of the antigen with lymphocytes, particularly T cells.

In general, protein antigens can generate both primary and secondary responses of antibody production.

The *primary response* is the antibody production after first-time exposure to an antigen. It commonly requires 5–10 days and has relatively low production of antibody, mainly of the IgM type, as opposed to the IgG type.

The *secondary response*, which occurs with the second exposure to the antigen, is quicker (1–3 days) and produces a greater amount of antibody, mainly the IgG type (sometimes IgA or IgE). The major reason for the enhanced secondary response is the proliferation of B memory cells during the primary response. The secondary response is also much more specific for the particular offending antigen than is the primary response.

Although protein antigens ultimately result in the production of antibodies, and antibodies are produced by plasma cells, which are B cell derivatives, T helper cells are necessary for this process, because T helper cells facilitate B memory cell proliferation and differentiation into plasma cells (**Figure 6-8**).

Carbohydrate antigens can stimulate a primary response, but they do not stimulate an enhanced secondary response. This is so because carbohydrate antigens do not act on T cells and thus, without the help of T cells, do not result in the production of memory cells. Thus, carbohydrate-induced immune responses tend to be shorter lived than protein-induced immune responses.

B memory cells are an expanded population of B cells that originate in an immune response and enable a quicker response to the same antigen in the future. B memory cells originate in lymphoid tissue germinal centers. They then circulate in the blood to establish positions in more distant body areas. T helper cells also form memory cells and are found in lymphoid tissues.

B lymphocytes can respond either to free-floating antigens or to antigens on other cells. Macrophages and other kinds of accessory cells (e.g. dendritic cells) release **interleukin-1** (**IL-1**), a protein that stimulates

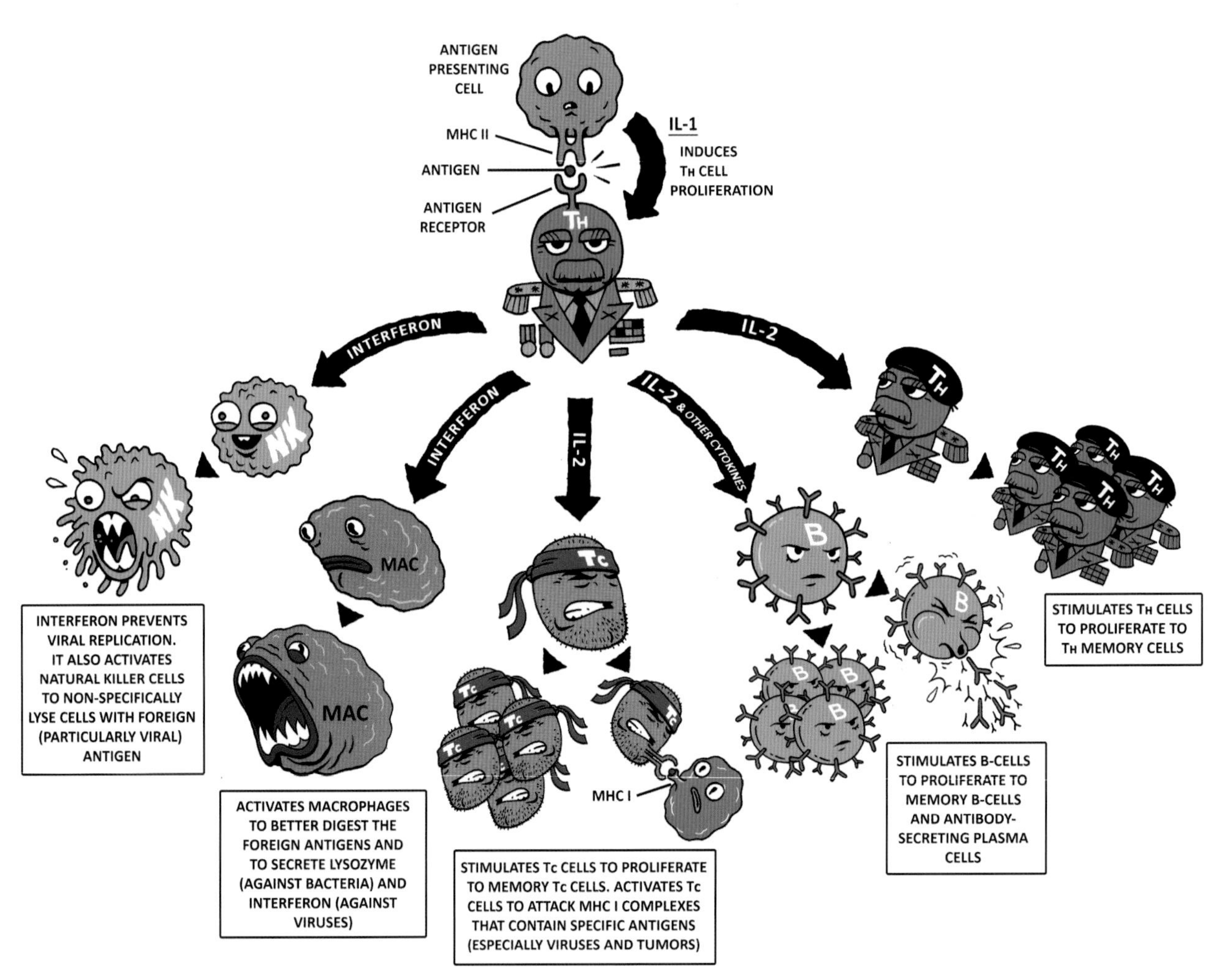

Figure 6-8. *T helper function. From Goldberg, S. Clinical Physiology Made Ridiculously Simple, Medmaster.*

T helper cells. T helper cells in turn produce cytokines that stimulate the proliferation and conversion of B lymphocytes to memory cells (**Figure 6-9**). Thus, the combination of antigen, cytokines from macrophages or other accessory cells, and cytokines from the T helper cells induces B cell proliferation.

Cell-mediated Immunity: T Lymphocytes

T cell lymphocytes originate in the bone marrow and mature in the **T**hymus gland. Rather than produce antibody, T cells contain cell-surface antibody-like molecules with which the T cell directly **T**ouches the antigen on antigen-presenting cells (so-called **cell-mediated immunity**). The antigen generally is presented to the T cell by another cell (such as a macrophage, spleen or lymph node dendritic cell, skin Langerhans cell, vascular endothelial cell, or B cell), stimulating the T cell to act (**Figure 6-9**).

Cell-mediated immunity differs from humoral immunity (the circulating antibody immunity of B cells) in that humoral antibody can be transferred from donor to host via cell-free plasma. Cell-mediated immunity, though, can only be transferred via T cells.

In general, free-floating antigens are dealt with through free-floating antibodies, whereas intracellular organisms, like viruses and certain bacteria, which cannot be reached by antibodies, are dealt with best through cell-mediated immunity, in which the cell containing the microbe is attacked. Both cell-mediated immunity and the antibodies of humoral immunity can act on antigens affixed to cell surfaces.

There are three major kinds of T cells: **T helper cells (Th)**, **T cytotoxic cells (Tc)**, and **T suppressor (regulatory) cells**.

1. **T helper (Th, CD4) cells**. When antigen presentation activates a Th cell, the Th cell secretes cytokines that stimulate B cell proliferation and antibody production. Not all B cells, though, require interaction with Th cells in order to produce antibodies.

Th cells are particularly sophisticated cells. In order to respond to the antigen in the first place, the antigen not only has to be properly introduced to the Th cell through another cell (commonly a macrophage, lymphoid tissue dendritic cell, skin Langerhans cell, vascular endothelial cell, or a B lymphocyte), but the etiquette of the introduction requires that the presenting cell be recognized as a member of the host's family, namely as a host cell. This family membership is evidenced by the presence on the presenting cell of a **major histocompatibility complex** (**MHC**, also called **HLA**, or "Human Leukocyte Antigen") that is common to the cells of the host (**Figures 6-8** and **6-9**). The MHC, while not an antibody, contains a number of amino acid sequences that resemble those in antibodies and render the cells of a particular host unique to the individual. Identical twins, though, have the identical MHC.

The receptor on the T helper cell thus is responsive to the combination of antigen and host-matching MHC complex on the presenting cell. In general, the presenting cell first phagocytoses the protein antigen and cleaves it to peptides in the presenting cell's lysosomes before extruding the modified (peptide) antigen, which attaches to the presenting cell's surface MHC complex. The presenting cell induces T helper cell proliferation (**Figure 6-8**). The T helper cell that has been affected produces cytokines that in turn stimulate further T cell and B cell proliferation and differentiation of B cells into plasma cells (**Figure 6-8**). Like B cell proliferation, T cell proliferation results in memory cells, but of the T cell variety. B and T memory cells can persist a lifetime.

The MHC on the antigen-presenting cell, in the normal immune response, is a specific type called **MHC II**.

MHC II is mainly found on macrophages and B lymphocytes, as well as dendritic cells and Langerhans cells. Another form of MHC, **MHC I**, is found more diffusely, on many kinds of cells. Both MHC I and MHC II are important in graft rejection. Th cells interact with those graft cells that contain MHC II receptors, whereas

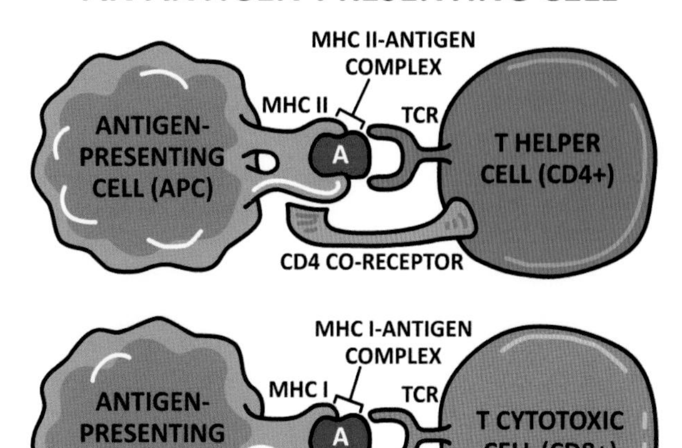

Figure 6-9. From Mahmoudi, M. Immunology Made Ridiculously Simple, Medmaster.

cytotoxic T cells (described below) interact with, and kill, the broad variety of foreign graft cells that contain MHC I receptors (**Figures 6-8** through **6-10**). Graft rejection may also be mediated through production of host antibodies against the graft.

It is believed that the particular MHC that an individual has may relate to the probability of developing particular autoimmune diseases.

T helper cells are quite "helpful." In addition to stimulating proliferation of B cells, T helper cells and T cytotoxic cells (discussed below), T helper cells also facilitate the removal of bacteria in still another way. Sometimes bacteria, although phagocytosed by macrophages, are not killed by the phagocytosing cells. Certain T helper cells, on recognizing bacterial antigen (in conjunction with MHC II complex) on the surface of the bacteria-filled macrophage, kill the phagocytosed bacteria. Cytokines are interesting in that they do not exhibit specificity for any particular bacterial or viral antigen, but nonetheless are an integral part of the specific immune response through their interaction with specific kinds of immune cells.

T helper cells also respond to foreign MHCs, such as might be introduced in a graft. Apparently, the T helper cells regard the foreign MHC as a "host MHC + antigen" combination and respond by stimulating T cytotoxic cells (see below), B cells and macrophages to attack foreign cells. The MHC thus is an important factor in transplantation rejection.

2. **Cytotoxic T cells (Tc cells, CD8)**. When antigen is presented to a B cell, the B cell attacks the antigen by converting into a plasma cell, which in turn produces antibody against the antigen. Cytotoxic T cells attack the antigen differently. The T cytotoxic cell is mainly concerned with defense against viruses, which are intracellular. It kills the cell that hosts the virus. Instead of producing antibody, the T cytotoxic cell directly combines with the degraded virus peptide antigen (previously degraded intracellularly), which is attached to the host's MHC I complex on the surface of the infected cell. In the case of graft rejection, the T cytotoxic cell considers the foreign MHC I (which is present on a wide variety of cells) to be a "host cell + virus" combination, and widely destroys the grafted cells.
3. **T suppressor (regulatory) cells** suppress the activation of the immune system. This helps promote tolerance to self, thus assisting in preventing autoimmune reactions. These cells also facilitate transplantation tolerance.

Figure 6-10. The T cytotoxic (CD8) cell on the target cell's MHC I complex. The cell targeted for death can be a foreign cell, or a host tumor cell with new antigens associated with the tumor cell's MHC I complex. Or the target could be a host cell that has been infected with a virus whose antigen attaches to the MHC I complex. There is a specific T cytotoxic cell for each type of viral antigen. T cytotoxic cells demonstrate a specificity for antigen, just as do the B and Th cell. As is the case for B cells, T cytotoxic cells proliferate to form memory cells with the help of T helper cells.

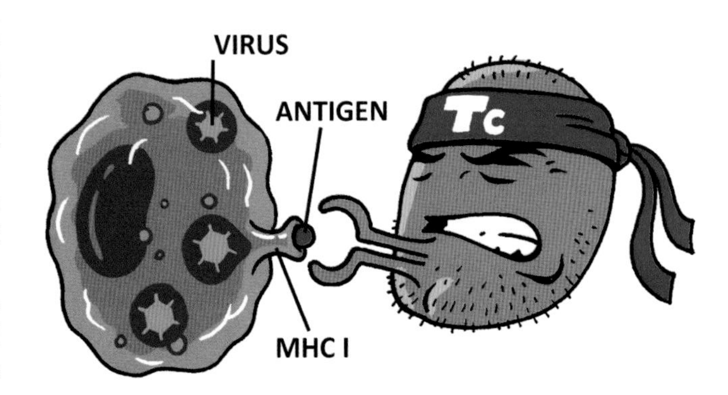

Figure 6-10. *T cytotoxic (T8) cell antigen interaction. From Goldberg, S. Clinical Physiology Made Ridiculously Simple, Medmaster.*

Thus, a T helper cell is first presented with antigen by an antigen-presenting cell, which also stimulates the T helper cell with cytokines (e.g. IL-1). The T helper cell then stimulates the production of more T helper cells, which stimulate proliferation of the T cytotoxic cells and the expression of the killer quality in T cytotoxic cells. The T-helper cell thus activates B cells, other T helper cells, and T cytotoxic cells. It really is a helper. Confusing? See **Figure 6-8** for clarification.

The various kinds of lymphocytes look alike but can be distinguished by chemical markers on their surfaces. Thus, among other markers, T helper cells contain the "CD4" marker, whereas T cytotoxic cells contain the "CD8" marker. Hence the names "T4" and "T8" lymphocytes for T helper and T cytotoxic cells. Special markers are also useful in distinguishing leukemic cells from normal cells.

Natural killer cell lymphocytes have been mentioned above as being more primitive forms that can kill microorganisms without the specificity of B and T cells. Natural killer cells do not require MHC on the target cells, but they kill virally infected cells in a way that is very similar to that used by T cytotoxic cells. NK cells do not require antigen to be expressed in the context of MHC; in fact, they appear to specifically target those cells that have reduced or lost their MHC proteins (**Figure 6-11).** This may come about as an adaptation induced by the virus (in an attempt to evade the T cytotoxic cell) or in a very abnormal cell (such as a tumor cell). Natural killer cells do not contain the typical B and T cell markers.

Immune Complexes

Figure 6-12. The reaction of antibody and antigen can result in large, precipitating complexes or small, soluble complexes, depending on the ratio of antibody to antigen. An equal ratio tends to produce a larger, precipitating complex.

Sometimes, all an antibody has to do is to react with an antigen and thereby neutralize it. Often, though, the antibody needs the help of complement or phagocytic cells to properly dispose of the offending antigen. The phagocytic cell engulfs the complex, whereas complement can solubilize complexes and also help attach them to phagocytic cells, Antigen-antibody complexes usually undergo phagocytosis and are removed.

If immune complexes are not removed, they may precipitate and damage normal tissue. Sometimes this results from defective phagocytosis. Damage may also occur through the unrestrained activation of complement by the antigen/antibody complex, and stimulation of neutrophils and macrophages to secrete lysosomal enzymes that damage normal tissue.

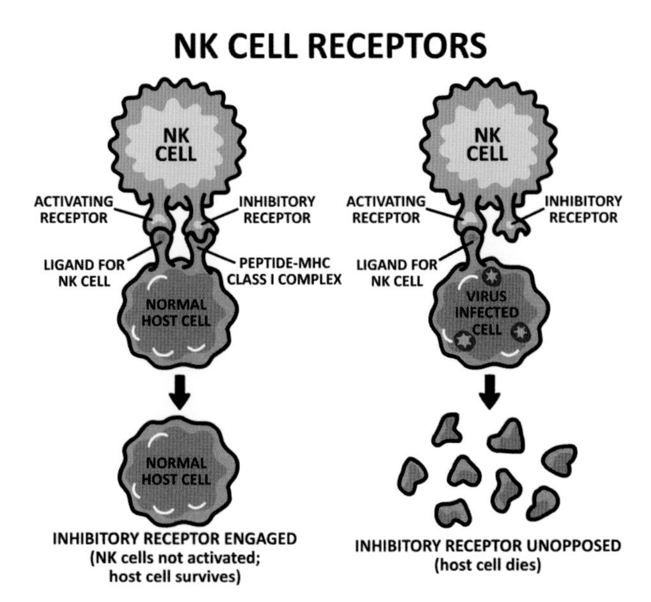

Figure 6-11. Natural killer cell action. From Mahmoudi, M. Immunology Made Ridiculously Simple, Medmaster.

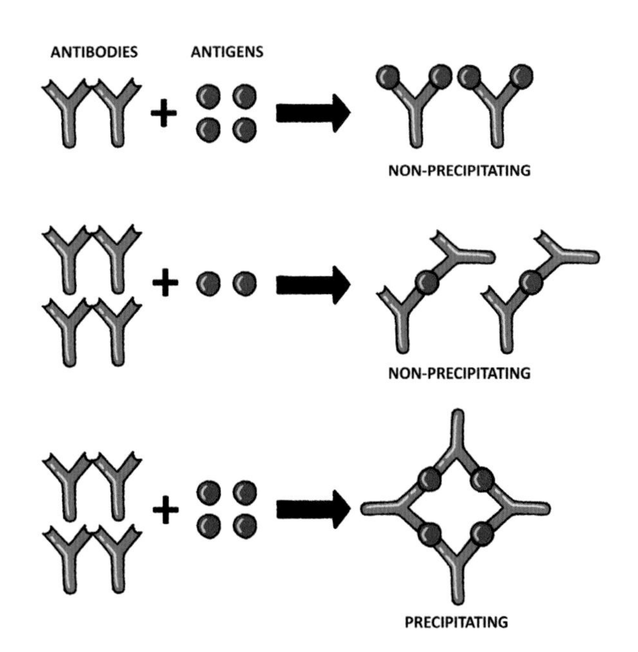

Figure 6-12. From Goldberg, S. Clinical Physiology Made Ridiculously Simple, Medmaster.

Natural and adaptive immune mechanisms do not act independently, but work together. Antibody-antigen complexes interact with complement and with nonspecific cells of natural immunity, such as macrophages, neutrophils, mast cells, or NK cells. For instance:

1. A phagocytic cell (**Figure 6-13**) may have:
   a. A receptor for bacterial antigens, through which bacteria are directly phagocytosed.
   b. A receptor for complement, through which the phagocytic cell deals with a complement-antigen complex, or a complement-antibody-antigen complex. Complement and phagocytic cells, although non-specific actors, are important in the elimination of antigen/antibody complexes.
   c. A receptor for the Fc portion of antibodies, through which the phagocytic cell deals with an antibody-antigen complex.
2. Antigen-antibody complexes can activate complement.
3. T cells, being culturally sophisticated, do not generally interact directly with free-floating antigens, but need to be properly introduced to them, commonly via macrophages that bring the antigens to the T cells.
4. Activated B and T cells activate the less specific macrophages and natural killer cells.

In view of the large number of B and T cell variations, the body cannot maintain a large supply of each one.

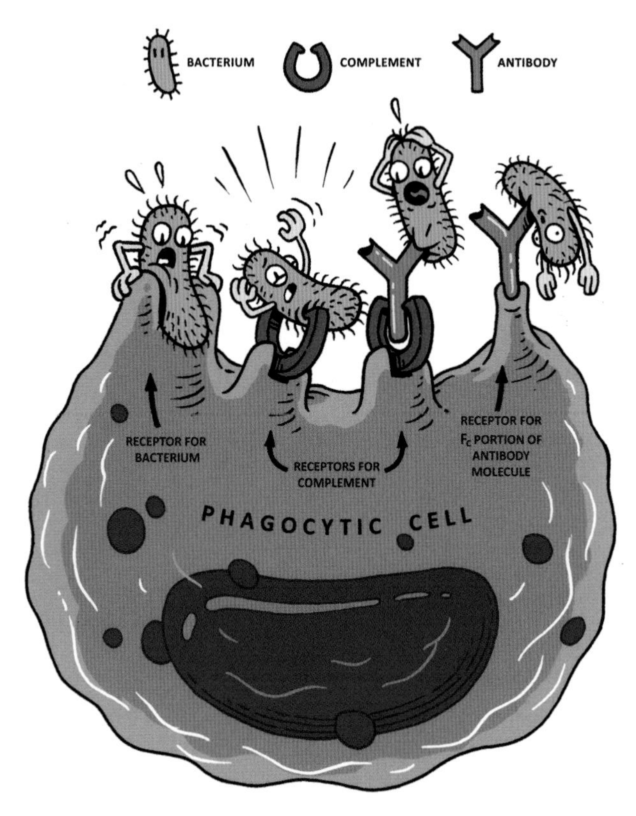

Figure 6-13. Phagocytic cell receptors. *From Goldberg, S. Clinical Physiology Made Ridiculously Simple. Medmaster.*

The ones that exist are strategically located throughout the body, particularly in lymph nodes, which drain the body, and within which there is a meshlike framework that can entrap antigens. When B lymphocytes (situated in lymph node germinal centers) are stimulated by specific antigens, those few lymphocytes divide, producing not only antibody-secreting plasma cells, but also memory cells. The memory cells persist and can more quickly mount an immune response should the same antigen be presented again.

Lymph node Th cells are mainly located near lymphatic follicles, where they are close to B cells and can exert their helper function. B and T lymphocytes may exit the lymph node via the lymphatic or blood circulations and end up in the general blood circulation, interstitial spaces, other lymph nodes, spleen or other organs. Plasma cells remain mainly in the lymph nodes and spleen, and produce antibody from there.

Lymph nodes, the spleen, and the thymus all have a blood supply, of course. Lymph nodes, though, have afferent and efferent lymphatics, enabling lymph nodes to deal mainly with lymph-borne antigens.

The spleen deals mainly with blood-borne antigens. Splenic arterioles end in vascular sinusoids of red pulp, which filter blood of nonimmunogenic foreign substances and old red blood cells. Near the red pulp are macrophages, dendritic cells, plasma cells and lymphocytes, in a region termed white pulp, which is like a large lymph node. Also, the spleen can act as a reservoir for blood cells, especially platelets.

The thymus has exiting lymphatics but no entering lymphatics.

When Things Go Wrong

A number of factors may interfere with a successful immune response:

1. Microorganisms may adapt and evade the immune response.
2. The immune response may be insufficient.
3. The immune response may be excessive, damaging the host in an attempt to fight the invader.

Microorganism Adaptation

Microorganisms may successfully adapt and defend themselves against the host response in the following ways:

- A bacterial capsule may prevent attachment to host macrophages.
- The bacterial cell wall may resist digestion and may contain endotoxins, toxic lipopolysaccharides found particularly in the cell walls of gram-negative organisms.
- The bacteria may secrete exotoxins, some of which can damage phagocytic cells; others (**aggressins**) alter the environment in a way that promotes spread of the bacteria through the tissues.
- The microorganism may demonstrate only weak antigenicity.
- New antigen variations in the microorganism may thwart the host's reaction just when the host has successfully developed an antibody response to the old antigen.

Insufficient Immune Response

The immune response may intrinsically be insufficient for a number of reasons, apart from a decrease in the number of white blood cells (**Figure 6-14**). Defects in the immune system may occur at any point in the immune mechanism—at the nonspecific level of complement, granulocytes and macrophages, or at the specific level of B and T lymphocytes and antibodies. Immune deficiency renders the patient susceptible to infections and, in some cases, tumors. The specific infection that develops may depend on which component of the immune system is defective. In some disorders, the immune system is not the only system affected, and diseases may occur with components other than infection and tumors.

At the level of natural immunity:

1. **Opsonization** (by complement and antibody) may be deficient. Phagocytosis, inflammation, and lysis of bacteria cannot occur effectively (**Figures 6-2** and **6-14**).
2. Neutrophils may lose the ability for **chemotaxis**, for adhering and moving along the endothelial surface and exiting into the tissues (e.g. **leukocyte adhesion deficiency**). The inflammatory infiltrates are then devoid of neutrophils, and pus doesn't form.
3. Neutrophils may lose the ability to **phagocytose** bacteria (**Chediak-Higashi syndrome**). Chediak-Higashi syndrome is a hereditary disorder that affects neutrophils, monocytes, lymphocytes, and other kinds of cells, resulting not only in infections, but hemorrhage (platelet involvement), neurologic impairment, and albinism.
4. Neutrophil lysosomes become non-functional, losing the ability to **lyse** bacteria (**chronic granulomatous disease; myeloperoxidase deficiency**).

Deficiency in phagocytic cells and complement will cause susceptibility to bacterial infections, since phagocytic cells and complement are the first-line defense against bacteria. Moreover, since complement is also important in the removal of immune complexes, patients with complement deficits may also develop diseases of immune complex deposition, with vasculitis, nephritis, and arthritis.

Individuals with T cell deficiencies are open to many infections but still may be able to handle microbes with polysaccharides as primary antigens, because the defense against polysaccharide antigens involves antibody that is T cell-independent. Defects in (B cell) humoral immunity render the patient more susceptible to pyogenic bacterial infections, or organisms for which B cells and their associated antibodies are the main defense. Defects in (T cell) cell-mediated immunity are more associated with defective responses to viruses (and viral-induced tumors), and organisms for which T cells are important in the defense, including intracellular bacteria (such as Tb), fungi, and protozoa. It must be remembered, though, that T cells are also important in B cell activation, and, therefore, T cell deficiencies can result in defects in humoral immunity.

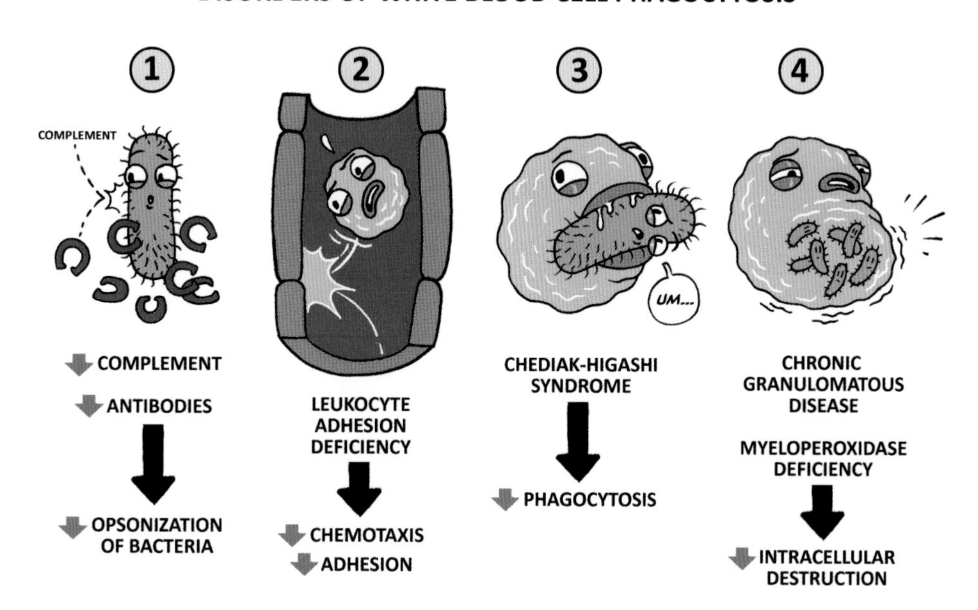

Figure 6-14. Modified from Berkowitz, A. Clinical Pathophysiology Made Ridiculously Simple. Medmaster.

Excessive Immune Response

The host response may be excessive, damaging the host (even during a normal immune response) with direct toxicity from non-specific components of the natural immunity response. The inflammation and cell lysis caused by excessive complement activation may be harmful. So also may be the toxic effects of excessive cellular and chemical interactions of phagocytic cells with normal tissues.

Complement may be excessively activated in resistant infections or in the continuing stimulation of an autoimmune reaction. Increased complement activity may also occur when feedback inhibitory components of the complement cascade are deficient. In **hereditary angioneurotic edema**, there are attacks of skin and mucosal (e.g. laryngeal) membrane edema from excessive complement activity. In **paroxysmal nocturnal hemoglobinuria**, red blood cells are susceptible to complement lysis because the normal red cell surface proteins that help prevent this are missing.

Damage may also result from hypersensitive components of the acquired immune response. Damage may be caused by:

- IgE (Type I hypersensitivity reactions)
- IgG and IgM (Type II hypersensitivity reactions)
- antibody-antigen complexes (Type III hypersensitivity reactions)
- T helper or T cytotoxic cells in cell-mediated immunity (Type IV hypersensitivity reactions)

Type I reactions (**Figure 6-15**) (e.g. pollen allergy) are acute reactions mediated by IgE. IgE attaches to Fc receptors on the blood basophils and tissue mast cells. When this occurs together with linking of the antigen with the IgE, the mast cell (or basophil) degranulates, releasing vasoactive substances. This can result in a spectrum from rashes to anaphylactic shock, with vasodilatory hypotension and bronchiolar spasm. Nonspecific cells cause the damage, but the condition is initiated by a specific antigen. Type I is the most common form of hypersensitivity. It can flair up within minutes.

Type II reactions are mediated by IgG or IgM antibodies. They may involve reactions against foreign antigens (e.g. infections, transplant rejection) or against self-antigens, with phagocytosis of the antigen-antibody complex (**Figure 6-16**) or direct cytotoxic reaction against the cell carrying the foreign antigen (**Figure 6-17**).

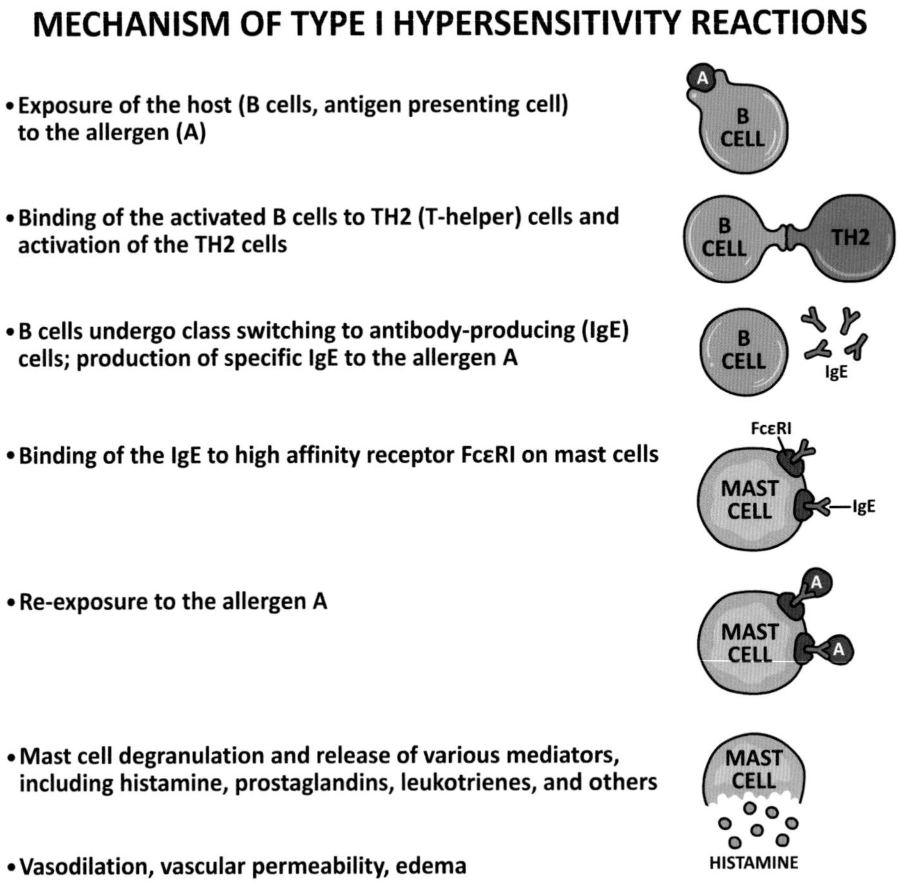

MECHANISM OF TYPE I HYPERSENSITIVITY REACTIONS

- Exposure of the host (B cells, antigen presenting cell) to the allergen (A)
- Binding of the activated B cells to TH2 (T-helper) cells and activation of the TH2 cells
- B cells undergo class switching to antibody-producing (IgE) cells; production of specific IgE to the allergen A
- Binding of the IgE to high affinity receptor FcεRI on mast cells
- Re-exposure to the allergen A
- Mast cell degranulation and release of various mediators, including histamine, prostaglandins, leukotrienes, and others
- Vasodilation, vascular permeability, edema

A
B CELL
B CELL
TH2
B CELL
IgE
FcεRI
MAST CELL
IgE
A
MAST CELL
A
MAST CELL
HISTAMINE

Figure 6-15. From Mahmoudi, M. Allergy and Asthma Made Ridiculously Simple, Medmaster.

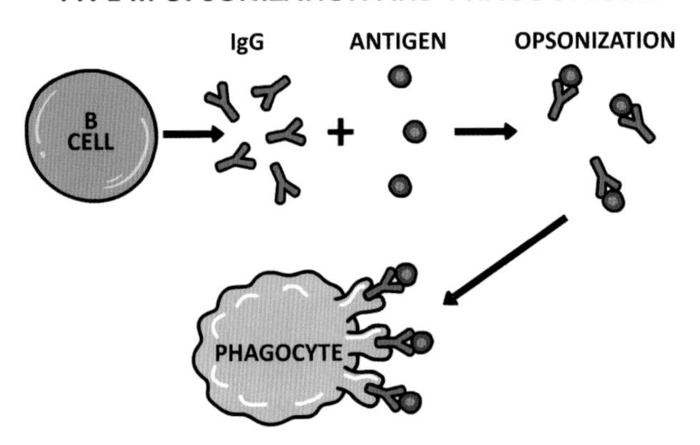

Figure 6-16. *From Mahmoudi, M. Allergy and Asthma Made Ridiculously Simple, Medmaster.*

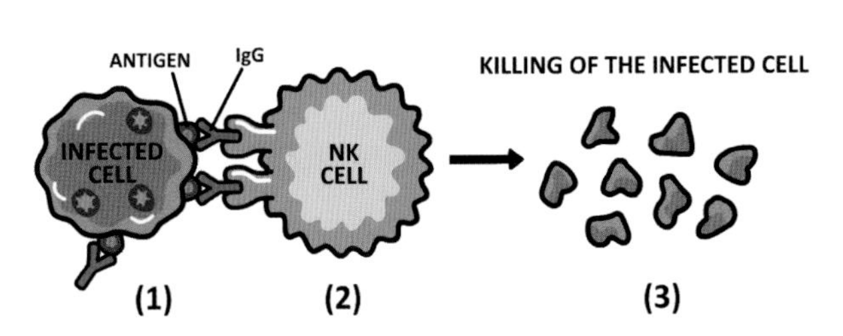

Figure 6-17. *From Mahmoudi, M. Allergy and Asthma Made Ridiculously Simple, Medmaster. NK = Natural Killer.*

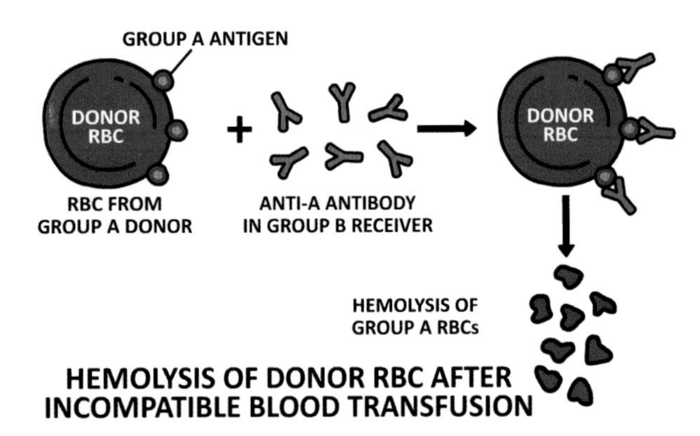

Figure 6-18. *From Mahmoudi, M. Immunology Made Ridiculously Simple, Medmaster.*

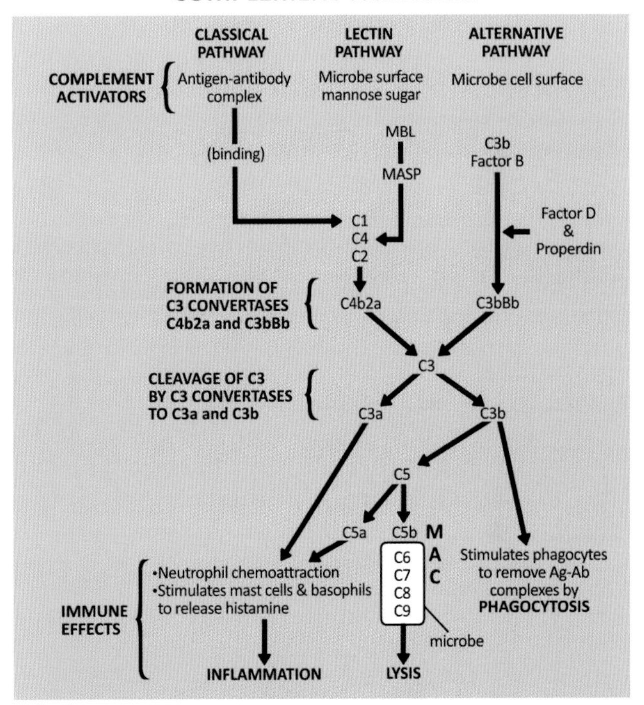

Figure 6-19. *Complement activators, by a variety of routes, result in inflammation, lysis, and phagocytosis. From Mahmoudi, M. Immunology Made Ridiculously Simple, Medmaster.*

Autoimmune hemolytic anemias (AIHA) are Type II antibody responses. Antibodies attach to RBC membrane antigens (**Figure 6-18**). The red blood cell is then hemolyzed either by *extracellular* phagocytosis (in the spleen or liver) or by direct hemolysis in the bloodstream (*intravascular* hemolysis). Although extravascular hemolysis is the most common AIHA, intravascular hemolysis is made more likely when *complement* is activated by the antigen-antibody complex, since complement has the special talent of drilling a hole directly into the red cell membrane. This causes hemolysis on the spot. Intravascular hemolysis is more likely when the antibody is IgM, which has a strong affinity for complement, particularly complement of the MAC (Membrane Attack Complex) type, which is excellent at drilling holes in cell membranes (**Figure 6-19**).

In autoimmune hemolysis, the antigen/antibody complex that is attached to the red cell membrane can be normal-occurring, like those in the A-B-O blood grouping, or can be foreign, such as a drug that causes an autoimmune hemolysis after attaching to the red cell. The harmful antibody against the drug attaches to the drug, and this combination activates the complement, which proceeds in drilling its hole into the red cell and hemolyzing it (**Figure 6-21**). Alternatively, the drug may remain free floating in the plasma, but generate an antibody response to itself, where the antibody cross-reacts with the normal antigens on the red cell (**Figure 6-20**).

In one condition, *Paroxysmal nocturnal hemoglobinuria (PNH)*, the complement alone causes the hemolysis without the antigen-antibody complex. The red cell

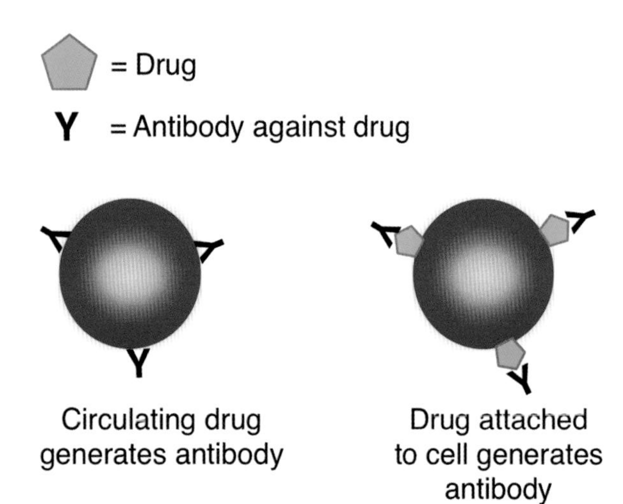

Figure 6-20. Two forms of drug-antibody interaction.

surface in this (non-inherited) gene mutation lacks a protein that normally protects against hemolytic attack by complement and is susceptible to hemolysis. PNH is not considered an autoimmune condition, since antigens and antibodies are not the precipitating factors, but rather a defect in the red cell membrane.

In **autoimmune thrombocytopenic purpura**, autoantibodies to platelet surface proteins result in thrombocytopenia.

In one form of **pernicious anemia**, autoantibodies to intrinsic factor deplete intrinsic factor, which is needed for vitamin B12 absorption. The symptoms are due to B12 deficiency.

Excessive production of monoclonal antibodies may be detected as a byproduct of certain lymphoproliferative disorders of plasma cells. In **Waldenstrom macroglobulinemia**, for instance, there is excess production of monoclonal IgM, coinciding with the anemia and bone marrow infiltration from abnormal, proliferating lymphocytes and plasma cells. The excess IgM may, among other things, cause blood hyperviscosity, resulting in sluggish blood flow, mental confusion, and retinal hemorrhages.

Multiple myeloma is a plasma cell tumor that invades bone. In addition to hypercalcemia from bone marrow destruction, there is increased production of monoclonal antibody (usually an IgG) and excess production of *antibody light chains.* The light chains filter through the kidney and are detectable on urine testing (**Bence-Jones protein**). Kidney damage may result from the hypercalcemia and Bence-Jones protein. Rarely, lymphatic malignancies may be associated with the excess production of *antibody heavy chains*.

Type III reactions (e.g. chronic glomerulonephritis, serum sickness, rheumatoid arthritis, Arthus reaction) are mediated by antigen/antibody complexes. Unphagocytosed antigen/antibody complexes may settle in tissues and excessively activate complement. The activated complement in turn activates neutrophils and macrophages to produce tissue-destructive lytic enzymes and phagocytose host cells. Neutrophils and macrophages can also be activated directly by immune complexes without the assistance of complement (**Figure 6-21**).

Immune complexes may cause widespread disease, but commonly they particularly affect blood vessels and highly filtering tissues, such as renal glomeruli (urine filtration) and synovial joints (synovial joint fluid). A common picture, therefore, is vasculitis, nephritis, and arthritis.

When the immune complex forms in the bloodstream, as in the infusion of foreign serum (**serum sickness**), the pathology may be widespread.

Examples of Type III reactions include:

- **Polyarteritis nodosa**. An immune complex of hepatitis B surface antigen (HBs Ag) and anti-HBs antibody results in diffuse arteritis.
- **Post-streptococcal glomerulonephritis**. An immune complex of antigens from the streptococcal wall with antibody attached to them results in glomerulonephritis.
- **Rheumatoid arthritis**. This condition involves autoantibodies (called *rheumatoid factors*) that likely contribute to the genesis of the disease, rather than just being a by-product of the disease.
- **Systemic lupus erythematosus**. Autoantibodies to DNA and nucleoproteins result in vasculitis, nephritis, and arthritis, among other things.

Type IV reactions (e.g. graft rejection, contact sensitivity, tuberculin response) are **cell-mediated reactions**. Damage may be caused by cytokines released

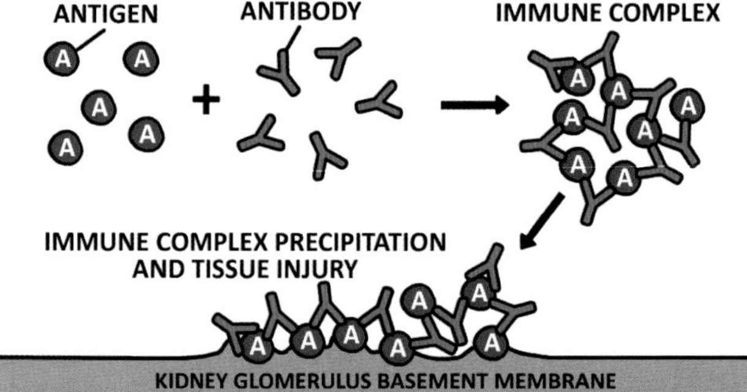

Figure 6-21. From Mahmoudi, M. Immunology Made Ridiculously Simple, Medmaster.

from T helper (T4) cells, by activated T cytotoxic (T8) cells, or by antibody production that is indirectly stimulated by T cell interaction with B cells (**Figure 6-22**). The killing-by-lysis action of the T cytotoxic (=CD8, T8) cells resembles that of the natural killer (NK) cells of natural immunity, except that NK cells do not have the specific affinity for certain antigens as is the case for CD8 cells.

Cell-mediated hypersensitivity reactions take longer than antibody-mediated reactions, often hours or days, the delay being due to the time required to mobilize cells through the T cell chain of interactions. **Contact sensitivity** is an example of this, in which skin Langerhans cells respond to antigen on the skin and initiate a T helper response. An eczematous rash may appear 1–2 days later. Such delayed reactions in skin tests have also been called "delayed hypersensitivity."

Tuberculin skin testing is a classic example of delayed hypersensitivity due to cell-mediated reactions. Tuberculin antigen is injected intradermally, and a skin reaction, in a person who has been exposed to tuberculosis, develops 48-72 hrs later in the form of an indurated reactive area on the skin. The reaction occurs through cell-mediated immunity.

In tuberculosis, cell-mediated immunity attempts to eliminate the tubercle bacillus by activating macrophages, which ingest the bacilli. If the ingested bacilli remain alive, the body may attempt a chronic inflammatory response, walling off the affected cells

TYPE IV HYPERSENSITIVITY

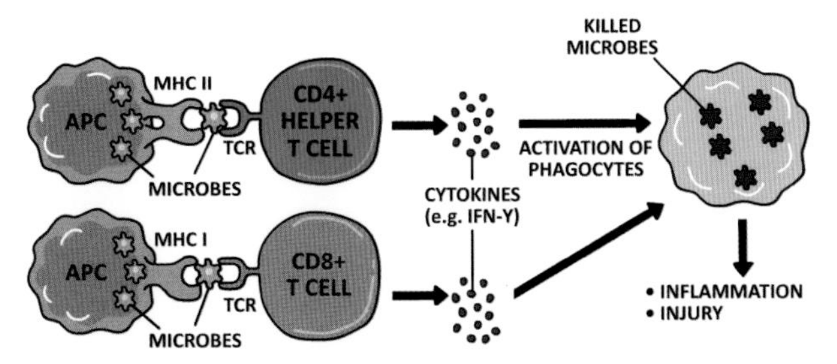

CD4+ HELPER T CELLS, CD8+ T CELLS AND CELL-MEDIATED HYPERSENSITIVITY REACTIONS

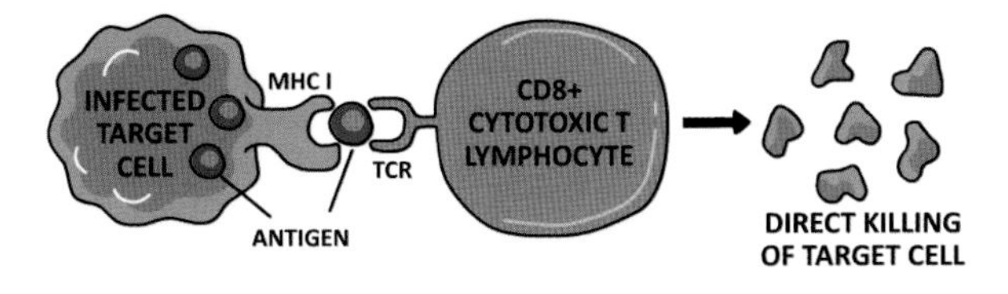

DIRECT KILLING OF INFECTED TARGET CELLS BY CD8+ CYTOTOXIC T LYMPHOCYTES

Figure 6-22. From Mahmoudi, M. Immunology Made Ridiculously Simple, Medmaster.

in damaging granulomas (**Figure 6-5**) in the lung and elsewhere. **Sarcoidosis** (nodules in lung, eyes, skin), **Crohn's disease** and **ulcerative colitis** (bowel), and **temporal arteritis** (arteries) are other examples of granulomatous diseases that are believed to involve cell-mediated immunity.

Figure 6-23 summarizes the hypersensitivity types.

Autoimmunity

Autoimmunity may be caused by either B or T cells, or both. Potential mechanisms include the following:

1. Breakdown of the body's ability to recognize self from non-self
2. The autoimmunity may not be against self antigens. Rather, a foreign antigen (e.g. virus or drug) attaches to the host cell. The immune response is against the foreign antigen, and in the process kills the host cell, too.
3. The foreign antigen may resemble a host antigen in some respect, and the immune response to the antigen cross-reacts and includes the host as well.
4. Certain antigens (e.g. bacterial lipopolysaccharide) can induce a generalized clonal proliferation of B cells (polyclonal activation), including the small population of self-reacting B cells that otherwise would be insufficient in number to mount an autoimmune response.
5. Antigens that appear later in development or are sequestered in the body and never participated in fetal self-tolerance may later generate an autoimmune reaction. For example, in severe ocular injury in which there is laceration of the choroid layer of the eye with permanent loss of vision, ophthalmologists may sometimes opt to remove the damaged eye, because leaving it in may lead to an autoimmune reaction against the opposite normal eye (**post-traumatic uveitis**) from antigens originally sequestered in the choroid but now released.
6. Homeostatic factors that keep the immune response in check may be deficient, as in some of the lymphoproliferative diseases.

FIGURE 6-23. HYPERSENSITIVITY TYPES

Reaction Type	Also Known As
Type I	Immediate hypersensitivity, anaphylactic, or IgE-mediated
Type II	Antibody (IgG or IgM)-mediated
Type III	Immune complex-mediated
Type IV	Cell-mediated or delayed type

AIDS

Defects in B or T cells can be secondary to chemotherapy, irradiation, malnutrition, hereditary factors, or invasive cancer. In the case of **AIDS** (**Acquired ImmunoDeficiency Syndrome**), T helper cells are destroyed or impaired by the AIDS virus (also called **HIV**, or **Human Immunodeficiency Virus**), which enters the T helper cell (also referred to as the CD4+ T cell, or T4 cell) via the cell's CD4 surface receptor. (The General directing the action in **Figure 6-9** is a "Four-star" General, for CD4.)

Antibodies first appear against the HIV virus about 3–20 weeks after exposure to the virus, which is transmitted through the exchange of body fluids (e.g. semen, through sexual contact; blood, through shared intravenous needles; breast milk or in utero transmission from an AIDS-carrying mother).

The first manifestation of HIV infections may be a flu-like illness, or the patient may be asymptomatic. A latent phase may then exist for up to 10 years or longer. Many patients in this time interval develop a syndrome consisting of lymphadenopathy, fever, weight loss, and diarrhea, as the number of T4 cells decrease. Eventually, most patients develop full-blown AIDS with profound loss of T4 cells. The ratio of T4 (T-helper) to T8 (T-cytotoxic) cells decreases, since T4 lymphocytes are destroyed by the virus. Normally a person has about a 2/1 ratio of T4 to T8 cells. This ratio may decrease to as much as 0.5 in patients with AIDS.

The patient with AIDS becomes susceptible to a variety of infections, some of which are otherwise rarely seen, such as the protozoan, *Pneumocystis pneumoniae*, a very common cause of death in AIDS. Other unusual protozoal infections (*Toxoplasma, Cryptosporidium*), bacterial infections (*Tb, Salmonella, Nocardia*), fungal infections (*Candida, Cryptococcus, Coccidioides, Histoplasma*), viral infections (*Cytomegalovirus, Herpes simplex, Varicella*), and helminth infections (*Strongyloides*) are also common. The patient may also develop **Kaposi sarcoma**, a rarely seen tumor.

B and T cell defects can also be hereditary. In the case of hereditary B cell deficiency, there may be a diffuse **hypogammaglobulinemia**; or the defect may be more selective, involving IgG, IgA, or IgM.

An example of a hereditary T cell deficit is the **DeGeorge syndrome**, in which there is failure of thymic development, and hence a deficiency in T cells. Patients with this condition, in addition to various congenital anomalies, also have impaired cell-mediated immunity, with normal or abnormal immunoglobulin levels.

In other hereditary conditions, there may be a combined deficiency of both B and T cells. In some of them, there may be multiple other problems. For instance, in the **Wiskott-Aldrich syndrome**, there are B and T cell defects, but also thrombocytopenia. In **ataxia telangiectasia**, there are, in addition to B and T cell defects, vascular malformations (**hemangiomas**) and neurologic problems.

Treatment: Immunosuppression and Immunostimulation

Immunosuppression is an important therapeutic modality that is used to diminish the harmful effects of immune hypersensitivity.

Disadvantages of immunosuppression include rendering the host susceptible to infection by damaging the immune system.

Approaches to immunosuppressive treatment include:

1. Deplete the lymphocyte population in general (antilymphocyte serum, antimitotic agents, irradiation, thoracic duct drainage).
2. Remove antibodies in general from plasma by plasma exchange.
3. Interfere with the function of cytokines in general by administering corticosteroids.
4. Interfere with the activities of lymphocytes. For example, **cyclosporin A** appears to interfere with T helper production of cytokines as well as with presentation of antigen to the T helper cell.
5. Specifically suppress the immune response to the particular antigen in the graft by:
    a. Administering specific antibody to block antigenic sites on the graft or on B and T cells.
    b. Administering specific antigen that is coupled to cytotoxic agents to destroy specific lymphocytes.
6. Induce tolerance to an immune response by administering massive doses of an antigen or small dosages given repeatedly. Clinical desensitization is an approach designed to decrease IgE levels in allergy by administering increasing quantities of the offending antigen in subcutaneous injections over a course of several weeks or months. The exact dose of antigen and how it is given may determine whether the response is stimulatory or suppressive to the immune system.

Sometimes a patient may display **anergy** (immunologically unresponsiveness) to an overwhelming infection. For example, in diffuse overwhelming tuberculosis, the patient may no longer exhibit a response to a tuberculin test.

Immunostimulation of the function of the immune system is possible through various methods:

1. **Passive immunization** through administration of antibody (e.g. tetanus, snakebite, hepatitis B) is useful to combat an existing infection.

2. **Active immunization** (vaccination) is useful to prevent future infection. Attenuated or killed microorganisms, or antigenic components of the microorganism are injected, thereby strengthening the immune system's future response should it encounter the live virulent strain.
3. Introduction of cells found in the immune response, e.g. grafting of bone marrow or thymus
4. The use of **adjuvants**, which are substances that enhance the antigenic response. Adjuvants may include substances not normally part of the normal immune response (e.g. oil emulsions, liposomes, as well as cytokines).

7

Leukemia

(Stephen Goldberg, MD and James Hoffman, MD)

Leukemia cells are cancerous white blood cells produced by the bone marrow. They are non-functioning. Rather than acting normal, leukemic cells behave like immature precursor cells (blasts) that keep on dividing, never mature, and don't die, but are non-the-less released into the blood stream.

Why is this a problem?

The problem is that unchecked proliferation of the abnormal cells in the marrow can crowd out the normal cells, so that there can be susceptibility to infection from the loss of normal white blood cells, anemia from loss of red blood cells, and a bleeding tendency from the loss of platelets. Leukemic cells can even invade the person's organs, including the brain and spinal cord.

Presenting symptoms are often vague, sometimes resembling the flu; they may include pain in the bones or joints, fever, fatigue, headaches, vomiting, shortness of breath, night sweats, enlarged lymph nodes, skin rashes and weight loss, as well as a feeling of abdominal fullness or pain (enlarged liver/spleen from leukemic cell invasion), easy bruising and bleeding, and seizures, depending on the type of leukemia and its degree of advancement.

Leukemias, like virtually all cancers, have a genetic basis. Genetic does not necessarily mean hereditary. Only about 5% of cancers are hereditary, and most leukemias do not run in families. Cancers arise from DNA mutations during the patient's lifetime, in somatic cells, not germ cells. While genetic, only a minority of leukemias are inherited (e.g. some cases of acute myeloid leukemia, myelodysplastic syndrome, and acute and chronic lymphoblastic leukemia). Others may arise from irradiation, certain chemicals, or spontaneously. Symptoms of leukemia can masquerade as infectious and rheumatologic diseases, orthopedic conditions, aplastic anemia, and other kinds of malignancies.

Types of Leukemia

Leukemia (the term "leuk" referring to white blood cells) can involve either the myeloid or lymphoid precursors of cells in the myeloid or lymphoid series in the bone marrow (**fig 2-1**). Hence, the terms myeloid (or myelogenous) leukemia and lymphocytic leukemia. Although there are many varieties of myeloid and lymphocytic leukemia, there are 4 primary categories to remember:

- Acute myeloid leukemia (AML)
- Chronic myeloid leukemia (CML)
- Acute lymphocytic leukemia (ALL)
- Chronic lymphocytic leukemia (CLL)

The term "acute" refers to the rapid progression of the leukemia, often requiring aggressive treatment. Chronic leukemias progress more slowly; the cells are more mature. Although they may look normal they aren't; they don't fight infection like normal cells. They live longer and crowd out normal cells in the marrow. In some cases, chronic leukemia may be asymptomatic for years. CLL in particular doesn't even require treatment in many cases.

Myeloid leukemias are cancers of the myeloid elements of the bone marrow, which are the precursors of the granulocytes (neutrophils, basophils, and eosinophils).

Lymphocytic leukemias are cancers of the precursors of the lymphocytes (B and T cells).

- Childhood leukemia is nearly always of the *acute* kind- most often ALL. *Chronic* leukemias rarely affect children.
- Acute lymphocytic leukemia (ALL) is the most common leukemia in young children (but can also occur in adults).
- Chronic lymphocytic leukemia is the most common leukemia in adults- and the most common of all blood cancers. It is often diagnosed after routine labs find an elevated white blood cell count- and the cells are found to be lymphocytes (often called 'smudge' cells on peripheral smear)

Chronic myeloid leukemia, which occurs primarily in adults, is always associated with the **Philadelphia chromosome** translocation (**Fig 7-1**).

Figure 7-1. The Philadelphia chromosome. An abnormal fused gene (red band) on chromosome 22 produces a new protein that promotes excessive production of white blood cells in chronic myelogenous leukemia.

Erythroleukemia arises from proliferation of immature red blood cell precursors. **Megakaryocytic leukemia** is a rare type that arises from megakaryocytes. Both of these are forms of AML (because red cell and platelet precursors are myeloid).

A unique form of AML is called APL (acute promyelocytic leukemia), which is caused by a translocation of the **15th** and **17th chromosomes** - t(15;17). This is unique in 3 ways:

1. Patients often present in **DIC** (**disseminated intravascular coagulation**).
2. Patients respond to a vitamin A derivative (**ATRA- all trans retinoic acid**), and to arsenic.
3. Because of these responses- the cure rates are >90%, as long as people can safely be treated through the DIC and managed through some early toxicities of the ATRA ('ATRA' syndrome = capillary leak, pulmonary edema -> managed with steroids.)

Kinds of Chronic Lymphocytic Leukemia include:

- **Prolymphytic leukemia (PLL):** The cells are immature B or T lymphocytes that grow and spread faster than the usual chronic lymphocytic leukemia.
- **Large granular lymphocyte (LGL) leukemia:** a rare form of chronic leukemia with features of T cells or natural killer (NK) cells. Most LGL leukemias are slow-growing, but a small number are more aggressive (they grow and spread quickly). Drugs that suppress the immune system may help, but the aggressive types are very hard to treat.
- **Hairy cell leukemia (HCL):** a type of B lymphocyte leukemia with hairy projections on its cell surface when viewed under the microscope. This is a highly treatable form.

Diagnosis of Leukemia

Diagnosis of leukemia may require:

- CBC: High white cell count with decreased red cells and platelets
- Blood smear: leukemic cells in blood
- Bone marrow biopsy

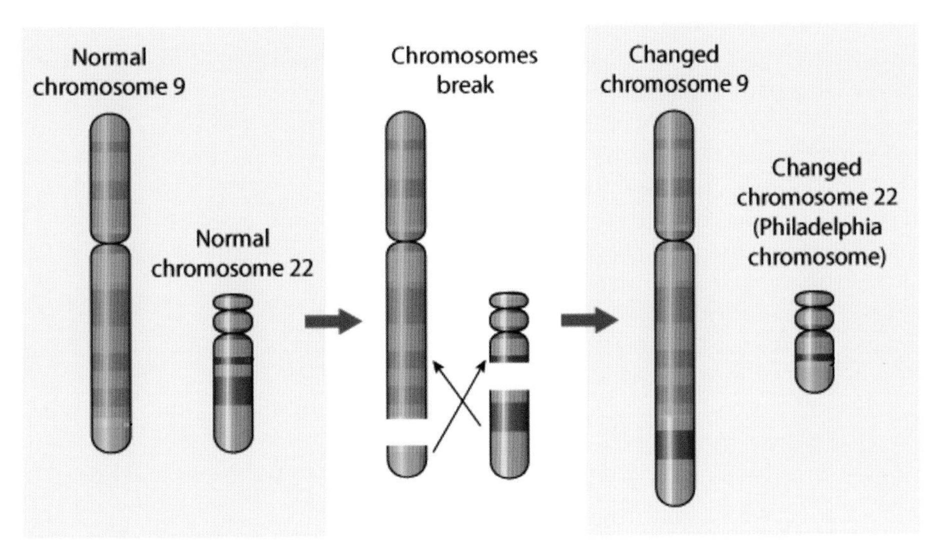

Figure 7-1. The Philadelphia chromosome.

- Spinal tap to see if leukemia has spread to the spinal cord/brain
- Imaging, e.g. x-ray, CT scan, MRI scan in some cases (not always)
- Genetic testing of the malignant white cells to confirm the kind of leukemia. This offers critical prognostic information, and can even identify novel targets for therapy.

Treatment of Leukemia

The treatment and cure rates for leukemia are continually improving. Acute lymphoblastic leukemia (ALL), the most common pediatric cancer, nearly always was fatal before the 1960's. The cure rate in children is now about 90%.

For AML, the cure rate is 65-70% in patients less than 20 years of age. For > 65 yrs, it remains a difficult disease, with long term disease control only achieved in 10-20% of people.

The main treatment approach for aggressive leukemia is chemotherapy to attack and kill rapidly dividing cells.

There are 2 general problems with chemotherapy in these patients. One is that normal cells that also divide rapidly may also be affected, with ensuing hair loss, mouth sores, nausea and vomiting, diarrhea or constipation; marrow suppression with increased risk of infection (decreased white cells), bleeding (decreased platelets) and anemic symptoms (decreased red cells).

The second problem relates to longer effects. Some chemotherapies can cause cardiac dysfunction (anthracyclines), many increase the risk of cancer later in life, and many cause infertility (men can sperm bank; for women, fertility preservation is more complicated).

It has long been a goal in oncology to avoid these strong chemotherapy drugs and offer a more targeted approach - sometimes referred to as 'precision medicine.' The treatment of leukemias does utilize these methods.

A more **targeted approach** narrows the treatment more directly to tumor cells. The approach depends on the cause of the genetic mutation. A normal cell has genes that act as tumor suppressors. A mutation to those genes may cause a **loss-of-function**, so that tumors are not suppressed. There are also **gain-of-function mutations**. I.e., Normal genes control the rate of cell growth and division and the normal time to die (**apoptosis**). A mutation may result in an abnormal protein that *increases* cell growth and division without dying, another cause of tumor proliferation. Some medications that are currently under study relate to the loss-of-function and gain-of-function mutations. Methylation of DNA and its associated histones can suppress gene function, while demethylation of DNA and histones, or acetylation of histones can increase gene function. Drugs that are either **methylators, demethylators, acetylators** or **deacetylators** are considerations for treatment.

- **Differentiation promoters**. Since leukemic cells act like immature cells that have never differentiated to mature cells, but keep on dividing and not dying, current trials are under way with drugs that can hasten the ability of cells to undergo differentiation. This is particularly used in APL, where the ATRA (all trans retinoic acid) forces differentiation.
- **Tyrosine kinase inhibitors (TKI)**. Tyrosine kinase (TK) is a cell enzyme that promotes cell proliferation, cell survival and migration, and when overexpressed, is associated with the development, progression, and recurrence of the cancer. The only TK provoked leukemia is **CML**. with the protein product of the t(9;22) being a TK that is constitutively activated (always on, like a light switch flipped up). *Imatinib*, a TKI, is arguably the most dramatic medicine in the history of hematology. Taking a gentle oral medication effectively cures most people with CML, and this was a disease that pre-2000 was the number one indication for stem cell transplant, and an often deadly disease. Newer TKIs have been developed for people that fail imatinib, and at this point, the biggest challenge in CML treatment is often getting patients to remember to take their pills!
    - In **CLL,** newer targeted drugs have also made a difference. While older chemotherapy drugs are still used, several more gentle 'precision' type approaches are commonly used. For example, *Ibrutinib* is an inhibitor of Bruton's Tyrosine Kinase (BTK) and is an effective daily pill, often used upfront.
- **Immunotherapy**. This includes
    a. ***Antibodies*** (e.g. the monoclonal antibody, *rituximab* and *alemtuzumab* for CLL) that target specific proteins on leukemic cells. Antibodies can even be attached to little chemo molecules that can then be brought directly to the cancer to kill them (e.g. *inotuzumab ozogamicin* for AML). These are called *antibody-drug conjugates (ADCs)*.
    b. ***CAR-T***. These are genetically enhanced T-cells that are more effective in fighting tumors. The immune cells are provided with an artificial gene that allows the cells to target and attack cancer cells. A CAR-T is already approved for relapsed ALL.
    c. **Vaccines** against specific antigens on the leukemic cell are being explored in current clinical trials. There have been many failures in this approach over time, unfortunately.
    d. **Allogeneic stem cell transplant**. Donor matched normal bone marrow stem cell transplants may be administered giving chemotherapy +/- radiotherapy to debulk the cancer and suppress

the immune system enough to accept the new stem cells. The new immune system (from sibling, or unrelated donor) can then recognize remaining leukemia cells (from the host) as 'foreign', and attack them. One of the perils of this approach is that the new immune system might find other parts of the host as foreign, and attack those parts as well. This is called *graft versus host disease*, and patients can suffer from rash, diarrhea, liver inflammation, etc.

- **Gene therapy**, the idea of directly removing the faulty gene or replacing it with a healthy one, is presently undergoing numerous trials. Marrow stem cells from the patient can be extracted, manipulated with gene therapy so as to disable the defective gene or substitute an effective one for the diseased one and then reintroduced into the patient. The reintroduced cells are those of the patient and are not subject to graft rejection.

Supportive care includes:

- **Antibiotics** for infection- often given prophylactically
- **Growth factors** (e.g. filgrastim) to increase WBC count after chemotherapy
- **Blood transfusion** may be used to provide healthy red and white cells and platelets.

Figure 7-2A and **7-2B** list drugs currently approved for the treatment of leukemia, per the National Cancer Institute.

FIGURE 7-2A. DRUGS APPROVED FOR LYMPHOCYTIC LEUKEMIA (NATIONAL CANCER INSTITUTE, 2019)

Acute Lymphocytic Leukemia	*Chronic Lymphocytic Leukemia*
• Arranon (Nelarabine)	• Acalabrutinib
• Asparaginase Erwinia chrysanthemi	• Alemtuzumab
• Asparlas (Calaspargase Pegol-mknl)	• Arzerra (Ofatumumab)
• Besponsa (Inotuzumab Ozogamicin)	• Bendamustine Hydrochloride
• Blinatumomab	• Bendeka (Bendamustine Hydrochloride)
• Blincyto (Blinatumomab)	• Calquence (Acalabrutinib)
• Calaspargase Pegol-mknl	• Campath (Alemtuzumab)
• Cerubidine (Daunorubicin Hydrochloride)	• Chlorambucil
• Clofarabine	• Copiktra (Duvelisib)
• Clolar (Clofarabine)	• Cyclophosphamide
• Cyclophosphamide	• Dexamethasone
• Cytarabine	• Duvelisib
• Dasatinib	• Fludarabine Phosphate
• Daunorubicin Hydrochloride	• Gazyva (Obinutuzumab)
• Dexamethasone	• Ibrutinib
• Doxorubicin Hydrochloride	• Idelalisib
• Erwinaze (Asparaginase Erwinia Chrysanthemi)	• Imbruvica (Ibrutinib)
• Gleevec (Imatinib Mesylate)	• Leukeran (Chlorambucil)
• Iclusig (Ponatinib Hydrochloride)	• Mechlorethamine Hydrochloride
• Inotuzumab Ozogamicin	• Mustargen (Mechlorethamine Hydrochloride)
• Imatinib Mesylate	• Obinutuzumab
• Kymriah (Tisagenlecleucel)	• Ofatumumab
• Marqibo (Vincristine Sulfate Liposome)	• Prednisone
• Mercaptopurine	• Rituxan (Rituximab)
• Methotrexate	• Rituxan Hycela (Rituximab and Hyaluronidase Human)
• Nelarabine	• Rituximab
• Oncaspar (Pegaspargase)	• Rituximab and Hyaluronidase Human
• Pegaspargase	• Treanda (Bendamustine Hydrochloride)
• Ponatinib Hydrochloride	• Truxima (Rituximab)
• Prednisone	• Venclexta (Venetoclax)
• Purinethol (Mercaptopurine)	• Venetoclax
• Purixan (Mercaptopurine)	• Zydelig (Idelalisib)
• Rubidomycin (Daunorubicin Hydrochloride)	
• Sprycel (Dasatinib)	
• Tisagenlecleucel	
• Trexall (Methotrexate)	
• Vincristine Sulfate	
• Vincristine Sulfate Liposome	

FIGURE 7-2B. DRUGS APPROVED FOR TREATMENT OF MYELOID LEUKEMIA (NATIONAL CANCER INSTITUTE, 2019)	
Acute Myeloid Leukemia	**Chronic Myeloid Leukemia**
• Arsenic Trioxide • Cerubidine (Daunorubicin Hydrochloride) • Cyclophosphamide • Cytarabine • Daunorubicin Hydrochloride • Daunorubicin Hydrochloride and Cytarabine Liposome • Daurismo (Glasdegib Maleate) • Dexamethasone • Doxorubicin Hydrochloride • Enasidenib Mesylate • Gemtuzumab Ozogamicin • Gilteritinib Fumarate • Glasdegib Maleate • Idamycin PFS (Idarubicin Hydrochloride) • Idarubicin Hydrochloride • Idhifa (Enasidenib Mesylate) • Ivosidenib • Midostaurin • Mitoxantrone Hydrochloride • Mylotarg (Gemtuzumab Ozogamicin) • Rubidomycin (Daunorubicin Hydrochloride) • Rydapt (Midostaurin) • Tabloid (Thioguanine) • Thioguanine • Tibsovo (Ivosidenib) • Trisenox (Arsenic Trioxide) • Venclexta (Venetoclax) • Venetoclax • Vincristine Sulfate • Vyxeos (Daunorubicin Hydrochloride and Cytarabine Liposome) • Xospata (Gilteritinib Fumarate)	• Bosulif (Bosutinib) • Bosutinib • Busulfan • Busulfex (Busulfan) • Cyclophosphamide • Cytarabine • Dasatinib • Dexamethasone • Gleevec (Imatinib Mesylate) • Hydrea (Hydroxyurea) • Hydroxyurea • Iclusig (Ponatinib Hydrochloride) • Imatinib Mesylate • Mechlorethamine Hydrochloride • Mustargen (Mechlorethamine Hydrochloride) • Myleran (Busulfan) • Nilotinib • Omacetaxine Mepesuccinate • Ponatinib Hydrochloride • Sprycel (Dasatinib) • Synribo (Omacetaxine Mepesuccinate) • Tasigna (Nilotinib)

8

Multiple Myeloma

(Stephen Goldberg, MD and James Hoffman, MD)

Multiple myeloma is a malignancy of plasma cells, which come from the family of B lymphocytes. Plasma cells are the most prolific antibody-producing cells, so patients with a malignancy clone of plasma cells, will usually have a single overproduced antibody detectable in their blood stream. It is by detecting and following this 'paraprotein (abnormal-protein)' that myeloma is most easily assessed.

A quick primer on measuring paraproteins: Antibodies have heavy and light chains. The primary heavy chain families are IgG, IgA, IgM and IgD. [Ig stands for immunoglobulin]. Each antibody either has an associated kappa (K) or lambda (L) light chain. So, an antibody can be IgGL, IgGK, IgAL, IgAK etc. Plasma cells can also secrete unattached lambda and kappa light chains- so we measure these 'free' light chains as well.

To evaluate and quantify a paraprotein, we:

1. measure IgG, IgA, and IgM levels (sometimes IgD, too)
2. check a serum protein electrophoresis; this is a tracing (pic below) that will show a monoclonal spike if an intact paraprotein is present (i.e. not free light chain only)
3. measure free kappa and lambda light chains
4. serum immunofixation - this will allow us to name the monoclonal spike (as IgGK, IgAL etc)
5. Consider 24 hour urine electrophoresis. This used to be the only way to quantify the free light chains, but with the advent of the serum free light chain assay it has been used less (much to the appreciation of the patients not having to carry around a day's worth of urine in a jug!)

Not every paraprotein equals myeloma. Clonal plasma cell diseases fall into several subtypes. To begin with, there is a spectrum of classic plasma cell disorders from MGUS (Monoclonal Gammopathy of Undetermined Significance) to Smoldering Multiple Myeloma (SMM) to MM.

MGUS is a low volume, benign appearing plasma cell disease. Its name literally means a single clone, antibody producing disorder, and we have no idea whether it

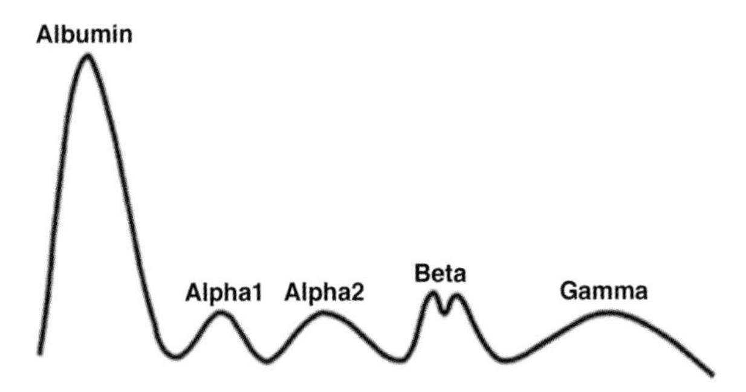

Figure 8-1. Normal serum protein electrophoresis (SPEP). ***Alpha-1 fraction****= alpha-1 antitrypsin, thyroid binding globulin.* ***Alpha-2 fraction****= ceruloplasm, haptoglobin.* ***Beta-1****= transferrin* ***Beta-2****= beta-lipoprotein [IgA, IgM, even IgG at times].* ***Between Beta and Gamma*** *= CRP, fibrinogen.* ***Gamma*** *= immunoglobulins.*

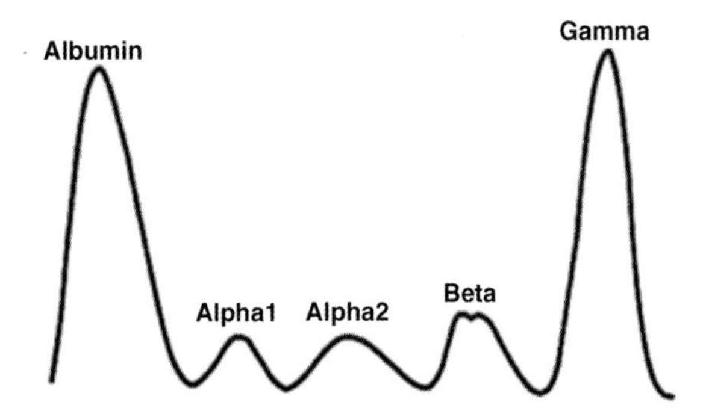

Figure 8-2. Typical Gamma M spike, meaning that there is a large amount of a very specific protein (i.e. produced by a clone) (Figs. 8-1 and 8-2 from NYU clinical website: https://www.clinicalcorrelations.org/2007/11/01/elevated-total-protein-and-the-interpretation-of-serum-protein-electrophoresis/ by James Hoffman, M.D.)

will ever be clinically significant. To qualify, you have to have plasma cells <10% in your marrow, and have a monoclonal spike <3gm. By definition, you can't be sick from it, or it loses its 'undetermined significance'.

Multiple myeloma is a proliferation of plasma cells, which develop from B lymphocytes. Myeloma tumors occur particularly in bone, and may cause anemia, thrombocytopenia, and leukopenia from crowding out of marrow cells by myeloma cells; painful bone destruction; hypercalcemia, fractures, and even spinal cord compression. The risk for infections is increased, since myeloma cells crowd out normal plasma cells and do not produce functioning antibody.

Since plasma cells produce one specific antibody, a characteristic laboratory finding is a spike in one particular kind of antibody (*monoclonal immunoglobulin*), as opposed to a general increase in antibodies. The abnormal antibody protein may injure the kidney and may be excreted in the urine (**Bence-Jones proteinuria**) where it can be detected.

Risk factors for multiple myeloma include:

- Increased age. Most people are diagnosed in their 60's.
- Male sex
- Black race. Multiple myeloma is about twice as prevalent in black people, as opposed to white people.
- Family history of multiple myeloma
- Diagnosis of a monoclonal gammopathy of undetermined significance (MGUS)

There are other conditions that are also marked by abnormal plasma cells:

- **Monoclonal gammopathy of undetermined significance (MGUS).** This form is by definition asymptomatic, and low volume, but may rarely develop into multiple myeloma (1% annual risk).
- **Smoldering multiple myeloma (SMM).** Also asymptomatic, this is a high volume plasma cell disorder (mspike >3gm or plasma cell % in the marrow >10) that often evolves into symptomatic myeloma (10% annual risk first 5 years, then 3% annually).
- **Solitary plasmacytoma.** This is a single malignant plasma cell tumor that may develop in bone, lung or other organs. It can sometimes be cured with focused radiation or surgery. Or it may develop into multiple myeloma, in which case the patient would need systemic chemotherapy.
- **Light chain amyloidosis.** The clonal plasma cells produce a pathologic antibody that misfolds into a waxy proteinaceous material (amyloid) that can cause organ damage (often to the heart, kidneys, GI track/liver or nerves).
- **Waldenstrom macroglobulinemia** is a type of B lymphoma of bone marrow and lymphatic tissue. It produces excess monoclonal IgM antibody. It results in more viscous blood, affecting blood flow through small blood vessels, with bleeding of the nose, retina, and gums along with an anemia.

Diagnosis of Multiple Myeloma

Diagnosis of multiple myeloma is made by

- detecting clonal plasma cells in the marrow, and generally a paraprotein (abnormal clonal antibody) in the blood
- end organ damage- "CRAB"= hyper**C**alcemia, **R**enal injury, **A**nemia and/or **B**one lesions

Workup includes

- bone marrow biopsy with plasma cell genetics analysis (FISH and cytogenetics). The FISH is a particularly important test because it establishes prognosis (there are higher and lower risk findings)
- skeletal imaging - at minimum X-rays, often CT, PET/CT and/or MRI
- paraprotein labs - SPEP, serum immunofixation, IgG, IgA, IgM, free light chain, Beta2microglobulin (B2M), CBC, CMP, LDH
- 24 hour urine for UPEP (urine protein electrophoresis)(less valuable given our ability to measure the free light chains in the serum). In the 'old days' the only way to measure the light chains was in the urine by UPEP- aka the 'Bence-Jones proteins'.

Treatment of Multiple Myeloma

Treatment options for multiple myeloma include (key classes **bolded)**:

- radiotherapy (for local problems or solitary plasmacytomas)
- **corticosteroids** (e.g. *usually dexamethasone*)
- immunomodulating agents (**IMIDs**) that affect specific aspects of myeloma metabolism (*lenalidomide, pomalidomide, thalidomide*)
- alkylating agents (*cyclophosphamide, melphalan*) to interfere with DNA synthesis
- **monoclonal antibodies** (*daratumumab, elotuzumab*)
- histone deacetylase inhibitor (*panobinostat*)- minimally beneficial
- **proteosome inhibitors** (*bortezomib, carfilzomib*) - kill myeloma cells by preventing proteosomes (a complex of proteases in cells that breaks down proteins) from recycling "garbage" proteins. This is particularly useful in myeloma, because the cells are so prolific making antibody proteins that they rely a lot on their 'garbage cans'.
- cellular immunotherapy (*CAR-T*) to manufacture a strong immune reaction against myeloma cells.
- plasmapheresis (for hyperviscosity)- most often used in Waldenstrom's (given the large size of IgM antibodies, hyperviscosity is more common).
- Supportive care (hemodialysis for renal failure, IV immunoglobulin for low gamma globulins, *denosumab* or *biphosphonates* to inhibit bone resorption)

For amyloidosis, treatment is similar, as the goal is to eliminate the bad plasma cells producing the pathologic antibody (that becomes the amyloid), so meds from the above classes are used.

9

Chronic Myeloproliferative Disorders (MPD) And Myelodysplastic Syndrome (MDS)

(Stephen Goldberg, MD and James Hoffman, MD)

Myeloproliferative malignancies, which mainly affect the older population, involve the proliferation of generally normal but too many cells in the myeloid line. We already discussed a chronic myeloproliferative disorder called CML (discussed in the leukemia section). There are 3 other forms to know (with some rarer subtypes). In these conditions there may be an excessive production of red blood cells, neutrophils, eosinophils, and/or platelets- remember, these are all myeloid.

There are 6 chronic types of myeloproliferative disorder, the most important 4 are **bolded**:

- **Polycythemia vera (PCV)**
- **Essential thrombocythemia (ET)**
- Chronic neutrophilic leukemia
- Chronic eosinophilic leukemia
- **Primary myelofibrosis (MF)**
- **Chronic myeloid leukemia (CML)**

Polycythemia Vera

Polycythemia vera, which only rarely runs in families, is an abnormal proliferation of normal red blood cells (and may be accompanied by too many white cells and platelets as well ('poly' + 'cythemia' = 'too many cells in the blood'). It is a rare form of blood cancer.

Diagnosis of Polycythemia Vera

Polycythemia vera is like a light switch always turned on in the marrow, with cells being produced even though the body doesn't need more, and may manifest as

- pain or a full feeling on the left upper abdomen due to splenomegaly from migration of the excess red cells to the spleen
- itching, particularly on bathing in warm water, which may have to do with the condition stimulating histamine release from mast cells
- reddish skin
- a high hematocrit (HCT >49% in men, >46% in women)
- bone marrow hypercellularity
- low erythropoietin (EPO) blood level. EPO is a hormone produced by the kidneys when the body needs more red cells. Here, the marrow has gone rogue, producing these at a rapid clip. The EPO levels drop very low as the body tries to send the message 'WE GOT ENOUGH'. Actually an EPO level can be valuable in the workup of a high HCT: a high EPO level means the body wants the high HCT (because of low oxygen levels from COPD or smoking for example) and a low EPO level means

the marrow is doing it on its own. i.e. that the patient has PCV.

- venous or arterial thrombosis, or hemorrhage. The extra RBCs can clog up the circulation and cause clots, and can also interfere with normal clotting and cause bleeding - kind of a 'rock and a hard place' scenario.
- low serum iron and ferritin, with increased TIBC, since iron is used up in the excess production of RBCs. Sometimes giving the patient iron seems like a good idea, but backfires as it helps the HCT get even higher.
- gout (from uric acid generation)
- genetic mutation in the JAK2 gene - this is *always* seen in PCV, and is the diagnostic test of choice in the right scenario (NOTE: This can also be seen in 50% of essential thrombocythemia and myeloproliferative cases- so having a JAK2 mutation means you have a myeloproliferative disorder for sure, just doesn't tell you which one.
- Splenomegaly

Treatment of Polycythemia Vera

Management of PCV is broken down into high risk and low risk patients.

Patients older than 65 or those with a history of a blood clot are high risk.

They get an aspirin a day to prevent blood clots, phlebotomy (removing blood) to a HCT <48 in men, <45 in women, and hydroxyurea (and gentle oral chemo pill) to suppress marrow production.

Patients <65, without a history of blood clots are low risk and simply get the aspirin a day and monitoring.

Of note, patients with PCV (and other MPDs) can evolve into AML (a chronic myeloid process becoming an acute one) with deadly consequences.

Essential Thrombocythemia

In **Essential thrombocythemia** there is an abnormal proliferation of **platelets (**>450 K/mcL), resulting from increased production of megakaryocytes in the bone marrow.

Diagnosis of Essential Thrombocythemia

- The patient may at first be asymptomatic, but may later report headaches, dizziness, chest pain, redness/numbness/tingling and burning of the hands and feet, with transient ischemic attacks, and stroke from excess blood clotting.
- There may be accompanying paradoxical bleeding, which may relate to abnormalities in platelet function.
- Consistently elevated platelet count.
- Bone marrow biopsy confirms an excess production of megakaryocytes.
- There may be enlargement of the spleen, which serves as a reservoir for platelets.

Thrombocythemia can be mild or deteriorate into myelofibrosis, bone marrow failure, or leukemia.

Treatment of Essential Thrombocythemia

Essential thrombocythemia is treated with:

- low-dose *aspirin* (sufficient if low risk- i.e. no clot history and <65 yo)
- *hydroxyurea* or *anagrelide* to reduce platelet production (added for high risk - i.e. clot history and >65 yo)

Chronic Neutrophilic Leukemia

In **chronic neutrophilic leukemia**, the bone marrow makes too many neutrophils.

Diagnosis of Chronic Neutrophilic Leukemia

The patient may experience fatigue, weight loss, night sweats, pruritus, bone pain, easy bruising, or gout (raised uric acid production), although most people are asymptomatic when the diagnosis is first made. There may be hepatosplenomegaly from infiltration by the abnormal cells. There is also a tendency for hemorrhage, particularly cerebral hemorrhage, which has been attributed either to a coinciding thrombocytopenia, platelet defect, or infiltration of cerebral vascular walls by neoplastic cells.

The diagnosis of CNL is confirmed by the presence of a sustained neutrophilia (>25 K/mcL, at least 80% neutrophils), enlarged liver and/or spleen, and by bone marrow exam.

Treatment of Chronic Neutrophilic Leukemia

Chronic neutrophilic leukemia can be treated with *hydroxyurea* (controls white cell counts and splenomegaly) and other oral chemotherapeutic agents, as well as *interferon-alpha*, an anti-mitotic agent. Stem cell transplantation can be curative in some cases.

Chronic Eosinophilic Leukemia

Chronic eosinophilic leukemia (hypereosinophilic syndrome) is a myeloproliferative disease of too many eosinophils.

Diagnosis of Chronic Eosinophilic Leukemia

The patient may experience swelling and itching around the eyes, lips, hands or feet. The eosinophils may infiltrate and damage the heart, lungs, intestine, and nervous system. The condition can progress to acute myeloid leukemia.

The diagnosis is based on clinical symptoms, a persistent eosinophilia >1.5 K/mcL, bone marrow exam, and gene analysis.

Treatment of Chronic Eosinophilic Leukemia

Treatment of eosinophilic leukemia may include

- chemotherapy (hydroxyuria, cyclophosphamide, corticosteroids, vincristine)
- removal of the spleen (which makes white blood cells)
- immunotherapy (e.g. interferon-alpha, an anti-mitotic agent)
- targeted therapy with tyrosine kinase inhibitors if certain mutations present
- stem cell transplant
- steroids

Myelofibrosis

In **myelofibrosis**, there is scarring/fibrosis of the bone marrow (perhaps due to cytokines released by abnormal megakaryocytes). This can occur out of the blue - or after years of another MPD (Essential thrombocythemia or Polycythemia vera). This makes sense as a marrow that is always overproducing cells is prone to scarring or fibrosis over time.

Diagnosis of Myelofibrosis

In myelofibrosis, the marrow is fibrotic and inhospitable to all the healthy cells. As such, we can see:

- anemia
- decreased numbers of white blood cells, increasing the risk of infection
- decreased numbers of platelets and a tendency to bleed.
- There are abnormal teardrop-shaped RBCs in the blood - as these cells have to squirm through a fibrotic maze to get to the bloodstream.
- There is fibrosis on bone marrow biopsy.
- confirmatory gene tests (50% with JAK2 mutation as mentioned above)
- There is frequent extramedullary hematopoiesis (blood production outside of the marrow, with splenomegaly and hepatomegaly. In the fetus, blood cells are made in those locations, so when the marrow gets fibrotic, the body reverts back to production in those sites.

Myelofibrosis may also progress to acute myeloid leukemia.

Treatment of Myelofibrosis

- Mild cases of myelofibrosis may not require treatment, but more serious cases may benefit from
- blood transfusions
- androgen therapy to enhance red blood cell production and improve anemia in some people
- allogeneic stem cell transplant (rarely, as these are often older patients)
- treating symptomatic splenomegaly if present (with low dose radiation or splenectomy).

Ruxolitinib is a JAK2 inhibitor. There was great hope that this would be a miracle targeted drug like imatinib for BCR-ABL gene mutation in CML. It is not. It works whether or not the JAK2 mutation is present by decreasing inflammation, shrinking the spleen and improving patients' energy.

Myelodysplasia ("myelo"= marrow, "dysplasia" = abnormal appearing)

Myelodysplasia, which predominantly affects the elderly, is a condition where the marrow is damaged. Myelodysplasia resembles a failing bone marrow on its way to developing **acute myeloid leukemia (AML)**. In fact, it used to be called "pre-leukemia." Myelodysplasia has fewer abnormal marrow and circulating blasts cells than AML; this represents a spectrum of disease. Myelodysplasia can range from a mild, indolent condition requiring no therapy, to an aggressive problem just short of AML.

Some myelodysplastic conditions can be caused by exposure to chemotherapy or radiation, or to toxic agents, such as benzene, lead, and pesticides. Often, the cause is unknown. Most often it is not inherited, but some inherited mutations may increase the tendency to develop myelodysplasia.

Both myeloproliferative disorders and myelodysplasia can progress to leukemia or myelofibrosis.

Diagnosis of Myelodysplasia

The patient may be asymptomatic or present with overall symptoms of cytopenia: general symptoms of anemia, infections, easy bruising and bleeding.

The physical exam may show hepatosplenomegaly and enlarged lymph nodes. The CBC may show:

- Anemia
- Thrombocytopenia
- Leukopenia with immature circulating neutrophils
- Atypical cells (basophilic stippling, Howell-Jolly bodies, nucleated red cells (**Figure 3-18**)
- Increased numbers of reticulocytes if there is delayed maturation of reticulocytes or a hemolytic component
- The bone marrow may show increased numbers of blasts and other immature cells, as well as fibrosis.

Treatment of Myelodysplasia

Treatment of myelodysplasia includes:

- *Active surveillance* (watchful waiting)
- *Blood transfusions*
- *Erythropoeitin-like agents* to stimulate red cell production
- *Granulocyte colony-stimulating factor* (e.g. filgrastim) to stimulate neutrophil production. Colony-stimulating factors are also used to treat neutropenia from myelosuppression in myeloid leukemia and aplastic anemia.
- *Bone marrow stem cell transplantation*
- *DNA methyltransferase inhibitors* (e.g. *azacitidine, decitabine*) to enhance the functioning of tumor suppressing genes, kill abnormal stem cells, and allow normal ones to replace them
- *Lenalidomide,* a drug with multiple mechanisms of action, including antitumor activity. It has been used successfully in patients with myelodysplasia caused by 5q chromosomal deletion. It is also used in multiple myeloma. It is also expensive.
- *Antibiotics* if needed
- *Iron chelation therapy* in some for iron overload from multiple transfusions

Systemic Mastocytosis

Systemic mastocytosis is a proliferation of mast cells, which accumulate in the body tissues. It sometimes precedes mast cell leukemia. The histamine release can cause severe allergic reactions. Mast cell proliferation may cause liver and spleen enlargement (mast cells accumulate in liver, spleen, bone marrow, small intestines, and skin), abdominal pain (hyperacidity from excess histamine release), bone marrow failure, skin lesions (pruritic red-brown macules), flushing, and fibrosis of the bone marrow, sometimes accompanied by polycythemia vera.

Diagnosis of Systemic Mastocytosis

The diagnostic features of systemic mastocytosis:

- The CBC may show anemia, thrombocytopenia and a high WBC count.
- There may be elevated blood *histamine* levels, elevated serum *trypase* (a marker of mast cell degranulation), and low blood albumin (an indicator of liver malfunction), along with elevated liver enzymes.
- Skin and bone marrow biopsy show clusters of mast cells.
- Bone scan may indicate lytic lesions, osteoporosis, or osteosclerosis.

Treatment of Systemic Mastocytosis

Treatment of systemic mastocytosis is largely symptomatic. It includes:

- Avoiding precipitating factors, such as alcohol, emotional stress, extreme temperatures, insect bites, and non-steroidal anti-inflammatory drugs, which may trigger histamine release and worsen an attack
- Antihistamines for itching and flushing
- Epinephrine for episodes of anaphylaxis
- Steroids to reduce the skin lesions
- Proton inhibitors for hyperacidity (from excess histamine, which stimulates stomach parietal cell acid secretion)
- Mast cell stabilizers (e.g. *chromolyn*) to block mast cell degranulation and release of histamine and other immune mediators
- Ultraviolet light to relieve skin symptoms

10

Lymphoma

(Stephen Goldberg, MD and James Hoffman, MD)

The type of leukemia depends on the type of white cell precursor that goes awry in the marrow. **Lymphomas** are tumors of lymphocytes. Why aren't they called leukemias? How does lymphoma differ from lymphocytic leukemia?

Leukemias arise in the bone marrow and are largely confined to the bone marrow and blood. Lymphomas are solid tumors that can start anywhere (e.g. bone marrow, thymus, spleen, tonsils, lymph nodes, digestive tract) that contains lymph nodes, but usually begin and settle in the lymph nodes. Lymphomas commonly present with enlarged lymph nodes in the neck, armpits, or groin.

The distinction, though, between lymphoma and lymphocytic leukemia is not that absolute, since lymphoma cells can be found in the circulation, and lymphatic leukemia cells can be found in the lymph nodes. For instance, in the later stages of *chronic lymphocytic leukemia*, cancer cells may appear in the lymph nodes, and the disease is then called *small lymphocytic lymphoma*. The symptoms of lymphoma and leukemia can resemble each other – fatigue, fever, night sweats, shortness of breath, and weight loss.

Hodgkin and Non-Hodgkin Lymphomas

There are about 70 different kinds of lymphomas, but they can be divided into **Hodgkin** and **non-Hodgkin** types:

- Non-Hodgkin is the most common lymphoma.
- While non-Hodgkin lymphomas may originate in lymph nodes of the neck, armpit or groin, Hodgkin lymphoma usually begins in the upper body (neck, chest or armpits)
- Hodgkin lymphomas are characterized by **Reed-Sternberg cells**, large cells with multiple nuclei that look like owl eyes (**Figure 10-1**).
- Patients with non-Hodgkin lymphoma usually are older than 55, while Hodgkin lymphoma usually occurs in younger patients, ages 20-34.
- Hodgkin lymphoma is usually diagnosed early and is more treatable, while non-Hodgkin lymphoma

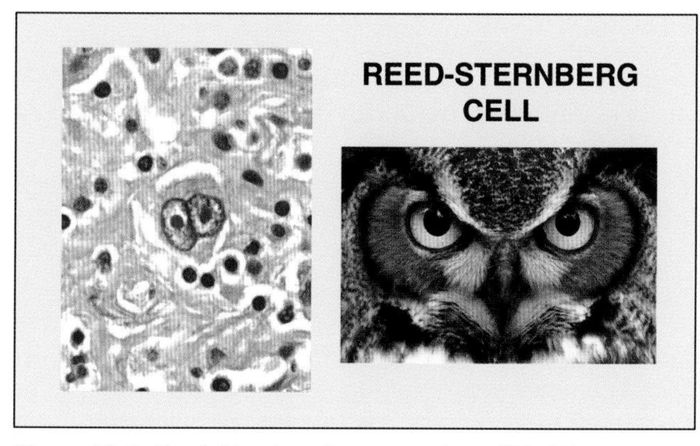

Figure 10-1. *Reed-Sternberg image courtesy of Dr. Dharam Ramnani, WebPathology.*

is usually more advanced before it is diagnosed. Hodgkin lymphoma, which used to have only a 10% 5-year survival rate, now has an over 90% survival rate for early stage and over 50% for stage 4 disease.
- Non-Hodgkin lymphomas can arise from B cells or T cells, usually from B-cells. Those arising from T cells can be found in the thymus. Reed-Sternberg cells in Hodgkin lymphoma originate from B cells. However, these are only a small part of the tumor, which contains T cells as well.

Lymphomas most often spread to the liver, bone marrow, or lungs.

Diagnosis of Lymphoma

- The patient may present with fever, weight loss, weakness and fatigue, itching, night sweats, and an asymmetrical, painless lymphadenopathy in the neck, underarm, and groin, with hepatosplenomegaly on exam.
- Lab tests may show a normocytic anemia. A low platelet and/or white blood cell count may indicate lymphoma presence in the marrow.
- Lymph node biopsy can provide more specific identification of the lymphoma type.
- A bone marrow biopsy may show lymphoma cells.
- Imaging tests to help determine the extent of spread include CT, MRI, and PET scans.

Treatment of Lymphoma

Lymphomas may be treated with chemotherapy, radiotherapy, immunotherapy, targeted therapy and, in some cases, stem-cell transplant.

A number of types of drugs are used that target cell division. The specific treatment, or combination of treatments, will depend on the specific lymphoma, its stage, and the patient's overall health.

- **Alkylating agents** interfere with DNA replication by reacting with purines in the DNA, causing the DNA strands to break, interfering with DNA replication, causing cell death (e.g. *cyclophosphamide*).
- **Antimetabolites** block certain of the metabolic pathways needed for DNA synthesis:
  a. They prevent ribonucleotides from converting to deoxyribonucleotides, *interfering with DNA replication*, e.g. *hydroxyurea.*
  b. Folate antagonists interfere with DNA replication by *preventing folate from being used to make DNA*, e.g. *methotrexate.*
  c. Pyrimidine nucleoside analogues incorporate into DNA, where they *inhibit DNA replication* (e.g. *cytosine arabinoside*).
  d. Purine nucleoside analogues incorporate into DNA, where they inhibit DNA replication (e.g. *fludarabine, mercaptopurine, azathioprine, bendamustine, clofarabine, pentostatin*).
- **Cytotoxic antibiotics** include:
  a. **Anthracyclines:** bind to and interfere with topoisomerase, an enzyme needed to uncoil DNA during replication, thereby interfering with DNA replication (e.g. *doxorubicin*)
  b. **Bleomycin:** creates superoxide radicals that cause breaks in DNA strands and interfere with DNA replication
  c. **Vincristine:** blocks microtubule formation, thereby interfering with the mitotic spindle during mitosis, killing the dividing cell
- **Targeted Drugs**: block specific proteins in cancer cells and are more specific to cancer cells than alkylating agents, antimetabolites, and cytotoxic antibiotics:
  a. **Tyrosine kinase inhibitors**. Tyrosine kinase is an enzyme that transfers a phosphate group to tyrosine in a protein. This process "turns on" the protein so that it is more active in promoting cell division. In some cancers, this enzyme is mutant and results in overactive cell division. Tyrosine kinase inhibitors (e.g. *imatinib*) prevent this overaction, leading to death of the cancer cell.
  b. **B-cell signaling pathway inhibitors**. *Bruton's tyrosine kinase activation* in B-cells has been correlated with the formation of certain B-cell malignancies. *Ibrutinib* and *acalbrutinib* are Bruton tyrosine kinase inhibitors used in the treatment of certain B-cell tumors. Other kinase inhibitors (e.g. *idelalisib* and *duvelisib)* are used in other B-cell cancers. Kinase inhibitors used in myelofibrosis and polycythemia vera include *crizotinib, midostaurin, gilteritinib, crenolanib* and others.
  c. **Isocitrate dehydrogenase inhibitors**. Isocitrate dehydrogenase 1 and 2, enzymes important in cell metabolism, have mutant forms in certain kinds of cancer and cause histone/DNA hypermethylation. Inhibitors of these enzymes are useful, particularly in acute myeloid leukemia (e.g. *ivosidenib, enasidenib).*
  d. **Proteosome inhibitors**. A proteosome is a protein complex that degrades proteins as a normal process in cellular metabolism. When such degradation does not occur properly, normal proteins may be degraded while

abnormal ones may not. For example, p.53 protein, which is important in cell division, may not be degraded when mutated, and stays at a high level, resulting in excessive cancer cell proliferation and tumor progression. Proteosome inhibitors (e.g. *bortezomid, ixazomib, carfilzomib)* may help guard against this and appear to act more specifically in certain cancer cells (e.g. myeloma, certain lymphomas) than in normal cells.

e. **Monoclonal antibodies** (e.g. *rituximab*) are particularly useful against B-cell malignancies by binding to CD20 protein on the surface of B cells, resulting in death of the B cell. Monoclonal antibodies to CD30 are useful in Hodgkin lymphoma. It is possible to attach toxins or radioactive isotopes to the antibodies, rendering them more effective. *Bilatumomab* is a bispecific T-cell engager monoclonal antibody. Namely, it contains two binding sites, one for T cells and one for the target B cells. It activates the T cell to recognize and destroy the target B cell.

f. **Epigenetic modifiers**. Mutant DNA can result in cancer by the nonfunctioning of tumor repressors or DNA repair mechanisms (loss-of-function mutations) or overfunctioning of genes that promote cell division (gain-of-function mutations). Epigenetic modifiers alter the functioning of genes without needing to change the DNA sequence. DNA methylation and histone deacetylation generally result in gene repression, while DNA demethylation and histone acetylation result in gene activation. Either the activation or deactivation of genes can be useful in cancer therapy, depending on the tumor.

g. **Interferon-alpha** inhibits mitosis during viral infection and inflammation. It has been useful in multiple myeloma, chronic myeloid leukemia, and myeloproliferative disorders.

h. **All-trans retinoic acid**, a derivative of vitamin A, promotes differentiation of cells in acute promyelocytic leukemia, stimulating production of more normal neutrophils.

i. **Asparaginase** has found use in acute myeloid leukemia, where the tumor cells need asparagine to make protein. Asparaginase breaks down asparagine, depriving the tumor cell of asparagine.

j. **Immunomodulatory drugs** include *thalidomide*, *lenalidomide*, and *pomalidomide*. Immunomodulatory drugs act in several ways, as anti-proliferation drugs and enhancers of immune-mediated antitumor activity.

k. **Steroids** have a role in lymphoid cancers and myeloma for their activity against lymphocytes.

l. **Platinum derivatives** kill cancer cells through changes in DNA structure.

m. **Arsenic trioxide** kills cells, inhibits proliferation, stimulates differentiation, and inhibits angiogenesis by interfering with a number of cellular pathways.

n. **Immune checkpoint inhibitors**, e.g. *pembrolizumab (Keytruda)* release the natural brake on the immune system, so that T cells can function more vigorously in recognizing and attacking tumors.

o. **CAR-T cells** are genetically enhanced T cells that are more effective in fighting tumors. The immune cells are provided with an artificial gene that allows the cells to target and attack cancer cells.

11

Splenomegaly

Normally, the spleen, which lies hidden under the ribcage in the left upper quadrant of the abdomen, is not palpable, except for its tip in newborns up to 3 months of age.

Functions of the Spleen

The spleen has several important functions in both red cell and white cell activity:

- It filters and removes old, damaged red blood cells.
- It acts as a reservoir of red blood cells for times of need, as in hemorrhage.
- It can produce red blood cells in time of need (*extramedullary hematopoiesis*).
- It is like a large lymph node. It synthesizes antibodies and removes antibody-coated red cells and bacteria.
- It is a large reservoir of monocytes and lymphocytes, which are released when needed to fight infection.
- It can produce white cells when needed.
- It is a large reservoir of platelets. The bone marrow does not store platelets.

"Too much of anything is not good." A spleen that is too large may *overfunction* by trapping and destroying normal red cells as well as abnormal ones, contributing to an anemia. It may trap an excess number of platelets, leading to thrombocytopenia and susceptibility to bleeding. It may also destroy white blood cells. It can outgrow its blood supply and experience infarction, which also contributes to a reduction in its normal function. The more a spleen grows, the more cells it entraps, and the more cells it entraps, the larger it grows—a vicious cycle.

Removal of the spleen predisposes to infection, as it is the largest lymph node in the body. Antibiotics and vaccinations are used for additional protection.

Causes of Splenomegaly

What would cause an enlarged spleen?

- Increasing the spleen's **workload** (as in hemolytic anemia, where the spleen has to work overtime removing damaged red blood cells) can result in splenomegaly.
- A lack of red blood cells (as may occur with bone marrow disease) can cause the spleen to respond by enlarging and producing its own red blood cells (**extramedullary hematopoiesis**), as it once did in the fetus.
- **Blockage of blood flow** exiting the spleen, as in sickle cell crisis (*splenic sequestration*), can cause blood to back up, resulting in splenomegaly and anemia from the sequestration and destruction of red cells. Backup of blood pressure into the spleen from portal hypertension (e.g. in cirrhosis) or heart failure or splenic, hepatic, or portal vein obstruction can also cause splenomegaly.
- Just as lymph nodes can enlarge during infection and inflammation, the spleen, which has its own lymphocytes and macrophages, may enlarge in

a variety of different infections (e.g. infectious mononucleosis, malaria) and inflammatory conditions (e.g. lupus).
- **Tumors or cysts** of the spleen, as well as infiltration by leukemic, lymphoma or other kinds of cells (e.g. in **Gaucher disease**), can also cause splenic enlargement.

Treatment of Splenomegaly

Treatment of splenomegaly depends on the specific medical problem:

- **Transfusions** to replace red blood cells in hemolytic anemia
- **Antibiotics** for bacterial infection that causes splenomegaly
- **Vaccination** (*H. influenzae, Streptococcus pneumoniae, Meningococcus*)
- **Splenectomy** for tumors, abscesses, cysts. In certain autoimmune diseases (e.g. autoimmune thrombocytopenic purpura, or autoimmune hemolysis), significant destruction of platelets and red cells occurs in the spleen, and splenectomy may be beneficial if the patient does not improve with medical management, e.g. steroids. In idiopathic splenomegaly, splenectomy may be indicated to relieve the pain and discomfort of the enlargement, and be on the safe side by removing any undetected malignancy.

12

Blood Transfusion

Transfusion and Blood Grouping

There are as many as 400 different blood group antigens, but ABO and Rh factors are the most commonly involved in transfusion reactions.

A-B-O Factors

A and/or B antigens occur on RBCs. If you don't have these antigens (type O blood), then you do have antibodies to them in your blood in a similar way as you would have antibodies to other foreign antigens that are "non-self." The appearance of significant quantities of antibodies may relate to exposure to small quantities of related foreign bacterial antigens in the gut after birth.

A and B antiGENS are called *agglutinoGENS*. The corresponding antibodies are called *agglutinins*. Individuals with type A, B, or AB blood contain red blood cells with A, B, or both A and B antigens respectively. Their blood will agglutinate if transfused into a type O individual, who has agglutinins to both A and B antigens. Type O individuals are "universal donors" since their red cells contain neither A nor B antigens and will not agglutinate when transfused (**Figure 12-1**). Nonetheless, it is better to avoid donating type O blood into a type A or B individual, since antibodies to A and B from the *donor* may to some degree cause hemolysis in the recipient.

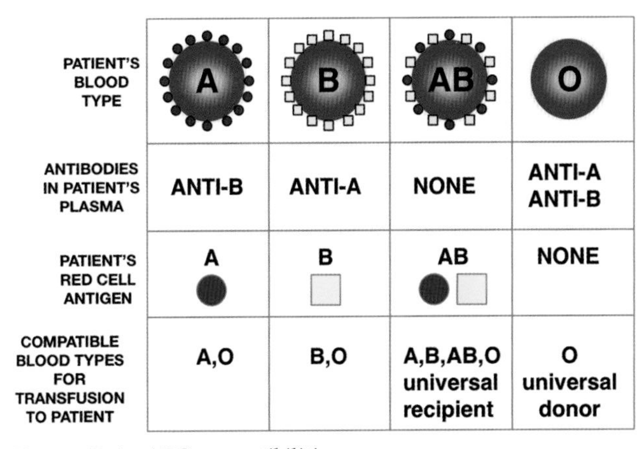

PATIENT'S BLOOD TYPE	A	B	AB	O
ANTIBODIES IN PATIENT'S PLASMA	ANTI-B	ANTI-A	NONE	ANTI-A ANTI-B
PATIENT'S RED CELL ANTIGEN	A	B	AB	NONE
COMPATIBLE BLOOD TYPES FOR TRANSFUSION TO PATIENT	A,O	B,O	A,B,AB,O universal recipient	O universal donor

Figure 12-1. ABO compatibilities.

Rh Factor

The ABO type of incompatibility occurs naturally. The Rh agglutinins do not, and there must be significant previous exposure to Rh antigens, generally by direct blood transfusion, before enough antibody can form. Hence, with the first pregnancy, with an Rh- mother and an Rh+ fetus, Rh incompatibility is not a problem. The second pregnancy with an Rh+ fetus may present a problem.

If the mother is Rh-, the baby will have 50% chance of being Rh+ if the father is heterozygous for Rh+, and a 100% chance if the father is homozygous.

People with the Rh antigen are called Rh positive. The main Rh antigen is called Type D antigen. There are other less potent ones. An Rh+ mother won't develop antibodies against Rh antigen, so her fetus will be unaffected, regardless of whether it is Rh+ or Rh-. If the mother is Rh-, then in response to the fetal cells of an Rh+ fetus, she will produce Rh antibodies.

Her antibody production will not occur in time to affect the first fetus but can affect the second pregnancy with an Rh+ fetus, since the antibodies are present for the second pregnancy. If the fetus is affected, this is **erythroblastosis fetalis**, a hemolytic condition of red blood cells. The term "erythroblastosis" refers to the nucleated RBC precursors in the blood that the infant develops to replace the RBCs damaged by the maternal antibodies. **Kernicterus** (damaging deposition of bilirubin in the brain, particularly in the basal ganglia) may arise from the hyperbilirubinemia that appears from the RBC breakdown. Even though the affected baby is Rh+, transfuse the baby with Rh- blood, because if you use Rh+ blood, it might react with Rh+ antibodies still circulating in the baby from the mother.

In order to prevent an Rh incompatibility reaction in pregnancy, the Rh- mother is administered anti-Rh factor antibody (anti-D) right after the first delivery of an Rh+ baby. This antibody combines with the fetal antigen and prevents the antigen from stimulating the mother's production of anti-Rh antibodies.

ABO hemolytic disease in the newborn, a disease separate from Rh incompatibility and usually mild, may occur during the first pregnancy. It is the most common maternofetal blood group incompatibility. The mother, who has type O blood, has anti-A or anti-B antibodies that cross the placenta and interact with the A and B antigens of the fetus.

A transfusion of Rh+ blood into an Rh- person (who has never contacted Rh+ blood before) causes no significant immediate reaction since it takes months for the antibody to develop to a significant degree. Within months, though, there may be a reaction after a second transfusion. Renal shutdown can occur from a transfusion reaction due to plugging of renal tubules with hemoglobin, in addition to renal vasoconstriction and circulatory shock that may arise from toxic products of the RBC breakdown.

Kinds of Blood Transfusion

Blood can be spun down and separated into plasma, buffy coat (white cells and platelets) and red cells. White cell transfusions are rarely given (There is little evidence that WBC transfusions reduce infection or death in patients with low or defective white cells). Instead, the clinician may administer *colony-stimulating factors* (e.g. filgrastim) to stimulate production of WBCs, mainly granulocytes like neutrophils. In some cases where the red blood cell count is low, the treatment may include *erythropoetin*, to stimulate the production of red cells in the patient, rather than risk an adverse reaction to transfusion. Transfusion reactions may include hemolysis, infection, and allergic reaction to WBCs, platelets, and other blood components.

Blood transfusions can consist of:

- **Whole blood** (with or without white cells). Fresh blood is best, since potassium is lost from red cells over time and in some cases could cause an adverse reaction.
- **Red cell concentrate** (used for most transfusions)
- **Whole fresh plasma** or plasma fractionated into **albumin**, **gammaglobulin**, specific **antiviral immunoglobulins**, or **anti-D immunoglobulin** to prevent hemolytic disease of the newborn, or **coagulation factors. Albumin** is used as a plasma expander when an osmotic effect is desirable (e.g. in hypoalbuminemia, hemorrhagic shock). Otherwise, fluid replacement with normal saline or Lactated Ringer's solution is usually used.
- **Cryoprecipitate of fresh plasma** (a source of fibrinogen)
- **Platelet concentrates** can be used in patients with thrombocytopenia or in patients who have dysfunctional platelets and serious hemorrhage.
- Both **red cell substitutes** and administration of **hematopoietic stem cells** are in an investigational stage.
- Granulocyte concentrates have sometimes been used in patients with severe neutropenia or severe infections not responding to antibiotics. Excluding the white cells from whole blood, however, is commonly done to avoid possible complications of a febrile transfusion reaction, formation of antibodies against HLA (Human Leukocyte Antigen), and transmission of certain infections (e.g. *Creutzfeldt-Jakob disease*, *cytomegalic inclusion virus*).

Autotransfusion (transfusion of the patient's own blood) is used in some cases where it is difficult to find a match for the patient's blood. Some of the patient's blood is removed prior to surgery (or blood lost during surgery is collected) and infused if needed during surgery.

Transplantation

Just as you can type RBC groups, you can tissue-type organs, examining for similarities in the **HLA (Human Leukocyte Antigen; MHC)** group of antigens, which are the most important antigens for tissue typing. The testing makes convenient use of lymphocytes, which contain the HLA antigens. If the patient's lymphocytes lyse when presented with a mixture of complement and antiserum against a specific HLA antigen, this suggests that the specific HLA antigen is present. HLA typing is thus useful to help maximize effectiveness in matching tissue donors and recipients. It is also used to help resolve paternity disputes. Specific HLA types are also useful in predicting the likelihood of developing certain infections, autoimmune diseases, inflammatory diseases, and cancers that are associated with particular HLA molecules.

Index

RAPID LEARNING AND RETENTION THROUGH THE MEDMASTER SERIES:

BASIC SCIENCES:
CLINICAL NEUROANATOMY MADE RIDICULOUSLY SIMPLE, by S. Goldberg
CLINICAL BIOCHEMISTRY MADE RIDICULOUSLY SIMPLE, by S. Goldberg
ORGANIC CHEMISTRY MADE RIDICULOUSLY SIMPLE, by G.A. Davis
CLINICAL ANATOMY MADE RIDICULOUSLY SIMPLE, by S. Goldberg and H. Ouellette
CLINICAL PHYSIOLOGY MADE RIDICULOUSLY SIMPLE, by S. Goldberg
CLINICAL MICROBIOLOGY MADE RIDICULOUSLY SIMPLE, by M. Gladwin, B. Trattler and C.S. Mahan
CLINICAL PHARMACOLOGY MADE RIDICULOUSLY SIMPLE, by J.M. Olson
ACID-BASE, FLUIDS, AND ELECTROLYTES MADE RIDICULOUSLY SIMPLE, by R. Preston
PATHOLOGY MADE RIDICULOUSLY SIMPLE, by A. Zaher
CLINICAL PATHOPHYSIOLOGY MADE RIDICULOUSLY SIMPLE, by A. Berkowitz
CLINICAL BIOSTATISTICS AND EPIDEMIOLOGY MADE RIDICULOUSLY SIMPLE, by A. Weaver and S. Goldberg
IS THE MIND IMMORTAL?: A Resolution of the Mind/Body Problem, by S. Goldberg
CONSCIOUSNESS MADE RIDICULOUSLY SIMPLE: A Serious Resolution of the Mind/Body Problem, by S. Goldberg
CLINICAL GENETICS MADE RIDICULOUSLY SIMPLE, by S. Goldberg

CLINICAL SCIENCES:
OPHTHALMOLOGY MADE RIDICULOUSLY SIMPLE, by S. Goldberg and W. Trattler
PSYCHIATRY MADE RIDICULOUSLY SIMPLE, by J. Nelson, W. Good and M. Ascher
CLINICAL PSYCHOPHARMACOLOGY MADE RIDICULOUSLY SIMPLE, by J. Preston and J. Johnson
BEHAVIORAL MEDICINE MADE RIDICULOUSLY SIMPLE, by F. Seitz and J. Carr
THE FOUR-MINUTE NEUROLOGIC EXAM, by S. Goldberg
CLINICAL RADIOLOGY MADE RIDICULOUSLY SIMPLE, by H. Ouellette and P. Tetreault
THE PRACTITIONER'S POCKET PAL: ULTRA RAPID MEDICAL REFERENCE, by J. Hancock
CLINICAL CARDIOLOGY MADE RIDICULOUSLY SIMPLE, by M.A. Chizner
CARDIAC DRUGS MADE RIDICULOUSLY SIMPLE, by M.A. Chizner and R.E. Chizner
CARDIAC PHYSICAL EXAM MADE RIDICULOUSLY SIMPLE, by M.A. Chizner
ECG INTERPRETATION MADE RIDICULOUSLY SIMPLE, by M.A. Chizner
ORTHOPEDICS MADE RIDICULOUSLY SIMPLE, by P. Tétreault and H. Ouellette
IMMUNOLOGY MADE RIDICULOUSLY SIMPLE, by M. Mahmoudi
ALLERGY AND ASTHMA MADE RIDICULOUSLY SIMPLE, by M. Mahmoudi
RHEUMATOLOGY MADE RIDICULOUSLY SIMPLE, by A.J. Brown
CRITICAL CARE AND HOSPITALIST MEDICINE MADE RIDICULOUSLY SIMPLE, by M. Donahoe and M.T. Gladwin
CLINICAL HEMATOLOGY MADE RIDICULOUSLY SIMPLE, by S. Goldberg
ARE YOU AFRAID OF SNAKES?: A Doctor's Exploration of Alternative Medicine, by C.S. Mahan
WAR AGAINST THE GERMS: EPIDEMICS, MICROORGANISMS, AND BIOWARFARE, by S. Goldberg

BOARD REVIEW:
USMLE STEP 1 MADE RIDICULOUSLY SIMPLE, by A. Carl
USMLE STEP 2 MADE RIDICULOUSLY SIMPLE, by A. Carl
USMLE STEP 3 MADE RIDICULOUSLY SIMPLE, by A. Carl
NCLEX-RN MADE RIDICULOUSLY SIMPLE, by A. Carl

OTHER:
MEDICAL SPANISH MADE RIDICULOUSLY SIMPLE, by T. Espinoza-Abrams
MED SCHOOL MADE RIDICULOUSLY SIMPLE, by S. Goldberg
MED'TOONS (260 humorous medical cartoons by the author) by S. Goldberg
THE STUTTERER'S APPRENTICE, by A. Splaver

For further information, see http://www.medmaster.net or Email: info@medmaster.net.